SYSTEMS CHANGE STRATEGIES IN EDUCATIONAL SETTINGS

New Vistas in Counseling Series
Series Editors—Garry Walz and Libby Benjamin

ERIC *Counseling and Personnel Services Information Center*

Structured Groups for Facilitating Development: Acquiring Life Skills, Resolving Life Themes, and Making Life Transitions, Volume 1
Drum, D. J., Ph.D. and Knott, J. E., Ph.D.

New Methods for Delivering Human Services, Volume 2
Jones, G. B., Ph.D., Dayton, C., Ph.D. and Gelatt, H. B., Ph.D.

Systems Change Strategies in Educational Settings, Volume 3
Arends, R. I., Ph.D. and Arends, J. H., Ph.D.

Counseling Older Persons: Careers, Retirement, Dying, Volume 4
Sinick, D., Ph.D.

Parent Education and Elementary Counseling, Volume 5
Lamb, J. and Lamb, W., Ph.D.

Counseling in Correctional Environments, Volume 6
Bennett, J. A., Ph.D., Rosenbaum, T. S., Ph.D. and McCullough, W. R., Ph.D.

Transcultural Counseling: Needs, Programs and Techniques, Volume 7
Walz, G., Ph.D., Benjamin, L., Ph.D., et al.

Career Resource Centers, Volume 8
Meerbach, J., Ph.D.

Behavior Modification Handbook for Helping Professionals, Volume 9
Mehrabian, A., Ph.D.

SYSTEMS CHANGE STRATEGIES IN EDUCATIONAL SETTINGS

Richard I. Arends
Jane H. Arends

Vol. 3 in the New Vistas in Counseling Series
Series Editors—Garry Walz and Libby Benjamin

 HUMAN SCIENCES PRESS

75 Fifth Avenue   3 Henrietta Street
NEW YORK, NY 10011 ● LONDON, WC2 8LU

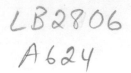

Library of Congress Catalog Number 77-22315
ISBN: 0-87705-310-3

Copyright © 1977 Human Sciences Press
72 Fifth Avenue, New York, N.Y. 10011

Printed in the United States of America
789 987654321

Library of Congress Cataloging in Publication Data

Arends, Richard.
    Systems change strategies in educational settings.

    (New vistas in counseling; v. 13)
    Includes bibliographies.
    1. School management and organization. 2. System analysis. 3. Group work in education. 4. Communication in education.
I. Arends, Jane H., joint author. II. Title. III. Series.
LB2806.A624   371.2′02   77-22315
ISBN 0-87705-310-3

TABLE OF CONTENTS

FOREWORD

When we commissioned Richard and Jane Arends to prepare a monograph for us on change strategies, we were mindful of the already abundant literature on the topic. In reviewing the resources on change, we had noted two factors that worked against their use by counselors and other helping professionals. First, very little, if any, of the literature was written with counselors in mind and hence would require extensive adaption by counselors to fit their role and circumstances. Second, much of the literature was technical and voluminous— hardly what time-pressed counselors needed to help them develop practical strategies for facilitating change in their systems.

It was our hope that the Arends could, by judicious sifting and sorting and by adding their own knowledge and flair, develop a monograph targeted for special use by counselors and helping professionals. We believe they have done just that. The authors are optimistic about the contributions that planned change can make in our schools and see a distinct role for the helping professional in the change process.

Counselors should be aware that change in their roles and functions increasingly must be the result of their own systematic efforts. They are likely to vary widely in their knowledgeability about change—ranging from an awakening awareness to sophisticated change agent skills. Fortunately, this monograph has something to offer for all, novice or experienced. Perhaps its greatest contribution is that it treats a diffuse and oftentimes confusing knowledge base with clarity and practicality.

Planned change is a young and emerging art, and ERIC is a rich vein of resources on change. We hope that this monograph will stimulate readers to utilize the wealth within the system, and will not only increase their knowledge about change but make them more prepared to include change agent functions within their role definitions. If so, both the system and the counselor are likely to benefit.

Garry R. Walz
Libby Benjamin

PREFACE

This monograph concentrates on the processes of change in a specific social system—the school—and presents specific strategies for promoting system change. It is addressed to school psychologists, counselors, social workers, and others responsible for providing counseling and pupil personnel services within school organizations. For some it may serve as a primer and help to satisfy a curiosity about the nature of system change; for others already involved in change projects, it can be used as a handbook with useful tools and strategies. Although our examples are specific to those who provide pupil personnel and psychological services in schools, we hope that the ideas and strategies are simultaneously helpful to others. Those who work in other kinds of systems, those who attempt to change systems in which they do not work all the time, and those who prepare people or groups to facilitate system change may also find the information in this monograph of interest.

We have written this monograph because we are basically optimistic. We belive it is possible to improve the human condition, and that good schools can provide one of our better vehicles for this improvement process. We also have a strong conviction that people in schools want to change and that helping professionals can and should take a more active role in promoting and facilitating change.

The monograph describes both the theoretical and practical aspects of producing system change. At times we write from an awareness of what researchers and theoreticians have said; at other times, we discuss the problems and issues from our own personal perspective, offering hints that have worked for us as we taught and consulted in schools. We do not claim that all of the suggestions will work for everyone in any situation, but we hope that they will at least stimulate thought and other writing.

We want our readers to know that we are not counselors, social workers, or school psychologists, but that we do care about these roles, the work associated with them, and the problems facing school systems today. We have worked in schools and have seen the frustration and joy that can be associated with school improvement and change. We have tried to bring about organizational change in schools and have studied the efforts of others who have been moving in the same direction. Since past opportunities to share with and learn from others have been very rewarding, this monograph has been written so that others might join our network.

The monograph is divided into five chapters; each chapter contains a related set of ideas. We know that there are other logical ways to structure and sequence these ideas, and we encourage readers to proceed in any direction that makes sense to them. In the first chapter we describe why we think schools need to change and suggest the part that those who provide guidance and pupil personnel services can play. The second chapter explains schools as social systems and describes some group processes through which work is accomplished in schools. Examples and tools appear in the third chapter, while the fourth includes guidelines. We suggest trends for the future in the final chapter and conclude with an annotated bibliography for those who wish to read further.

We thank all of those who helped make this monograph possible: Dr. Libby Benjamin who extended the invitation; Dr. Gordon Lindbloom and Fran Page who told us what they would want to read; and our colleagues, Ruth Emory, Dr. Martha Harris, Dr. Charles Jung, Phil Kessinger, Dr. Matthew Miles, and Rene Pino, who read the manuscript and provided so many helpful suggestions. A special thanks to our good friend, Lynn O'Brien, who typed the final draft during her Christmas vacation.

Richard I. Arends

Jane H. Arends

This chapter describes the needs and the possibilities for change in schools. We begin by citing evidence that the roles of counselors and school psychologists are changing to allow them to be a major influence for change in schools. Then we point to the need for change in schools and to the goals for the kinds of change that we have in mind. In a description of our perspective on change, we describe how purposeless change, mindless change, hit-and-miss change, and imposed change can be avoided.

Chapter 1

CHANGE IN SCHOOLS

We discern a trend in the counseling and psychological services profession away from the preventive approach to the mental health of individuals toward a systems orientation. This emerging role is frequently described in the literature and we believe is likely to take hold. Singer, Whiton, and Fried argue that "any psychological approach to education based only upon either psychopathology or a theory of individuals is inadequate" (1970, p. 173). They propose a model and role definition for those who provide psychological services based on (1) "working in a school . . . to help the school carry out its educational mission"; and (2) providing assistance to people in schools regarding the "dynamic aspects of education, the psycho-social phenomena which affect learning" (p. 174).

More recently, Murray and Schmuck described how the role-fragmentation of school counselors has reduced their effectiveness. They also cite evidence regarding the failure of the mental health model. They predict a role change for counselors and school psychologists that would "include moving away from attempts to increase the mental health of individual students through counseling toward attempts to improve the climate of the school organization by consulting with all members of the school" (1972, p. 99).

In the future we think that even more counselors and school psychologists will become aware of the limited potential and inefficiency of individual change and will try instead to bring about system change. They will intervene in subsystems, changing norms and roles that hinder meaningful and productive work. Medway (1975) describes the role we have in mind for counselors and school psychologists as that of "internally accountable socioeducational specialists" (p. 21). He describes these helpers as "behavioral scientists adept at reducing and containing the incidence of maladaptive social behaviors through indirect and consultation methods, organization development and systems analysis, and environmental restructuring" (p. 21). Medway says socioeducational specialists, working collaboratively with their school personnel clients, would do the following kinds of things: "(1) diagnosing interpersonal and educational aspects of the school culture and delineating present and projected service needs, (2) developing and implementing interventions in response to these needs, (3) evaluating planned changes, and if necessary, (4) devising alternative strategies" (p. 21).

Medway also recommends that counselors and school psychologists take a "synergistic" view of school organizations. This view implies that socioeducational specialists

"concern themselves with individual clients (e.g., students, staff, administrators) as social clients (e.g., classroom culture, faculty relationships, system communication processes, community influence, etc.) Helping individuals fulfill their needs, clarify their values and aspirations, gain a sense of accomplishment and acquire desired skills must be coordinated with efforts to help groups move from rival to teammate relationships" (p. 21.)

Meyers (1973), in describing a consultation model for school psychological services, suggests that counselors and school pyschologists diagnose and intervene at four levels: (1) direct service to the child such as testing and recommending treatment approaches, (2) indirect service to the child by coaching teachers in ways they might change classroom procedures to help the child, (3) direct service to teachers such as helping them to understand and plan ways to alleviate school-home problems, and (4) service to the school system. Meyers describes this final level as follows.

"Service to the school system, the fourth level of consultation in this model, can be differentiated from the three preceding levels since change in children or individual teachers is not the consultant's primary goal. On the contrary, the primary focus would be to change the behavior of various subgroups within the school. Two approaches to intervention can be conceptualized within this level of consultation. First, the consultant might take an active role in developing innovations, and this could be accomplished at an administrative level or through in service training. Some current examples might be the implementation of open classrooms, the implementation of modified grading procedures, or the restructuring of special education. A second approach to intervention would be to improve the general functioning of the school, as the consultant might help to facilitate communication between subgroups of administrators, teachers, or both, and the goal would be improved problem solving" (p. 11).

Having suggested this trend for the participation of counselors and school psychologists in the process of system change, the remainder of this chapter specifies our view of change in schools. It is organized in three sections: the first describes why we think schools need to change; the second outlines important goals for change; and the third explicates our perspective on change by reporting some research on what makes some change efforts succeed while others fail.

SCHOOLS NEED TO CHANGE

Certainly no one will deny that schools have changed in the past decades. Go back to the high school you graduated from and, without much trouble, you will observe changes in policies, modes of behavior, and topics of conversation. Even though few people would claim to have predicted accurately what schools in the 1970s would be like, we think everyone would agree that change has been fairly universal.

But what actually has changed about schools? Some would argue that the changes have been only on the surface—that schools perform the same basic societal functions that they did long ago. Even Sarason, a psychologist who has studied the problems of change in education, seriously commented that, "the more things change, the more they remain the same" (1971, p. 7). Others who have looked for new and potentially exciting practices in schools have come away equally disappointed (see Goodlad, 1970 and 1975; Silberman, 1970). Lortie summarizes the situation in this way: "If there has been a revolution . . . it has occurred in people's expectations for schools, not in practice; the gap between the possible and the actual has become an issue" (1975, p. 218). We share his observations and believe that the gap between "what is" and what many people think "ought to be" is the major reason why schools so badly need to change.

However, in addition to assuming that change is needed, we base this monograph on the assumption that resources for change exist. Some may challenge this assumption by pointing to the fact that it has become much harder in recent years to obtain money from the federal government and private foundations. The years when educators could opportunistically grab for abundant financial resources seem to be over.

Others may say that the critical resource of a large, young, and innovative staff is missing. As the decline in student enrollment causes reductions in the teaching force and as the mean age of teachers increases, others find an excuse they need for not changing.

There is still another common challenge to the assumption that resources exist. Even when the money and staff are sufficient, you can always count on some people to say "the time is not right." They point to pervasive and sometimes violent confrontations over teacher salaries, desegragation, or textbooks as the reason why "yesterday" or "later" would be better than now. Fearing additional confrontation, they find it much easier to wish for what used to be than to dream of what could be.

Yet we maintain that the resources are there in the form of ideas and human energy that typically go untapped. We have seen many examples of people learning to take great delight in sharing what they know and can do and in dreaming up new ways to organize themselves so that they can use all their resources to meet the demands of some current problem. We have seen whole staffs reach out to parents for their ideas and welcome parent volunteers as active participants in the instruction of students. We have seen committees dream up alternatives to *Robert's Rules* so that their meetings are more productive. We have known many schools where schedules and requirements have been changed so that students could tutor or counsel other students. And, in many of these instances, we have seen counselors and school psychologists serving as helping persons or change agents.

Certainly in our minds, the resources of helping professionals all too often go to waste. Efforts to change the behaviors, attitudes, or understandings of individuals are too typically ineffective and inefficient means for bringing about system change. If we could capture only part of the energy that goes into one-on-one counseling sessions, encounter groups, personal growth labs, treatment groups for underachievers or truants, etc., we would have more than enough to bring about important and lasting change in systems and organizations.

The list could go on and on, but we hope that we have made our point. The resources for effecting system change are there. The need is there. All that remains is for people to find their resources and bring them to bear on important problems. No one can do that for anyone else. Basically, that is what taking the chance to facilitate change is all about.

GOALS FOR CHANGE

From our value perspective, we believe schools need to become more satisfying places in which people work and learn. They need to become more humanistic settings where learners and teachers develop and reach their full potential for living in a changing world. Involvement, humanism, and relevancy need to become more of a part of the fabric from which schools are made.

But even if some schools were to change enough to meet expectations like ours, we would not be satisfied.

We believe that future generations will expect different things from schools. We believe what is most needed now is to change the ways that schools change so that *future* gaps between what schools will do and what people will want them to do can be closed.

Miles (1967) talked about the change goals we seek when he suggested the needs for: (1) increased internal interdependence and collaboration; (2) added adaptation mechanisms and skills; (3) stronger data-based, inquiring stances toward change, and (4) continuing commitment to organizational and personal growth and development. He deplored the isolation of individual educators within schools as making it "very difficult for school personnel to secure help and support from one another, to develop adequate solutions to educational problems, and to diffuse these inventions to others" (1967, p. 24). He spoke to the need for adaptation of schools to the communities they serve and emphasized the simultaneous need to insure that school-community transactions be of high quality. Similarly, he proposed that increasing the number of sensory and feedback loops within school organizations would transmit the information needed for short-run operations and for longer-term change. He concluded by saying that the orientation cannot be toward specific change projects alone, but toward institutionalizing the research and development change functions within schools.

In another article, Miles and Lake (1967) posited the following criteria by which movement toward institutionalized change capacity could be gauged: "(1) the effectiveness of problem solving by all groups in the organization; (2) high self-sustaining motivations of members to accomplish the goals of their groups; (3) effective linkage between the goals of one group and those connected to it; (4) work procedures and structures which are technically sound (will reach group and system goals); (5) a working climate which supports effective job performance; and (6) reward systems which facilitate cooperative and collaborative effort among groups and individuals, so that influence is exerted on the basis of actual competence and knowledge, rather than through organizational status as such" (Miles and Lake, 1967, p. 82).

Therefore, it is toward strengthening the capacity of school systems to change themselves that we think change efforts should be directed. And it is in moving toward these kinds of changes that we believe counselors and school psychologists must and will play a part.

A PERSPECTIVE ON CHANGE

Many people have tried to change schools in the past. Some have been reasonably successful while others have met dismal failure. Researchers who have tried to find out why some efforts work while others do not have generated numerous explanations. The four we discuss in the following paragraphs are only some that appear repeatedly as "keys to success" or "roadblocks to avoid." We assume that these four—purposeless change, mindless change, hit-and-miss change, and imposed change—are common enough to provide leverage in virtually any system.

The four topics reflect our choice of a perspective for system change that will be elaborated more fully in Chapter II. According to categories[1] of perspectives devised by Havelock (1970), our perspective falls more or less clearly among problem-solving models. This perspective emphasizes the social-psychological needs and motivations of people in the system to be changed.

Purposeless Change

Results of a very extensive study of educational change by the Rand Corporation (Berman and McLaughlin, 1975) suggest that change projects can be initiated in either opportunistic or problem-solving ways. Projects with opportunistic start-ups are those designed primarily to take advantage of funding or the availability of helpers. Projects that start out with a problem-solving focus, on the other hand, are initiated to help meet a specific need. They are designed and planned in ways that take account of present and expected future realities.

Projects studied by Rand that were generated essentially by opportunism were characterized throughout by a lack of interest and commitment on the part of participants. As a result, participants were often indifferent to project activities and outcomes. Little in the way of serious change was ever attempted—or occurred.

On the other hand, projects that started out with a problem-solving thrust were more likely to result in change. Whenever the project addressed goals that were important to teachers, principals, parents, and students, commitment and consequent action were there from the beginning.

One difficulty that must be overcome if purposeful and problem-solving change is to occur is the general lack of clarity and agreement about indicators of educational success. Since educators cannot talk about the rate of production or profits made last year, they often attempt change without specifying how they will know when they have been successful. Without clearly defined purposes and a technology for measuring progress toward goals, many change interventions stray off target and no one can tell until it is too late. Purposeful change thus provides leverage because it conserves human energy for movement toward important targets.

Mindless Change

Even when people clearly know what it is they want to change, they often move through the change effort as through a fog. Silberman (1970, pp. 10-11) defined this leverage most succinctly.

> "What makes change possible, moreover, is that what is mostly wrong with the public schools is due not to banality or indifference or stupidity, but to mindlessness. If they (teachers, principals, and superintendents) make a botch of it, and an uncomfortably large number do, it is because it simply never occurs to more than a handful to ask why they are doing what they are doing."

Silberman's words reinforce need for purposeful change and also speak eloquently to the need for consciousness about change processes. The primary importance of the Rand Study lies in the help it provides for thinking clearly about the *ways* in which educational change projects succeed or fail in meeting their objectives.

Rational planning, data collection, documentation, debriefing, and evaluation are all necessary if people would know *how* their purposes are being achieved. But because these activities are rare, important leverage for system change can be found through careful self-reflection and thoughtful self-analysis. Those who would change a system can often find one key to success by examining what happened in prior change efforts in their system.

Hit-and-Miss Change

Imagine, if you will, an orchestra where only the violins are tuned, or a football team where only the quarterbacks practice. Unless the violins play solo or unless no one cares that the ball always drops to the ground in midfield, this sounds ridiculous. Yet educational change efforts are often designed in this way. For instance, some people expect teachers to start teaming without taking time to figure out how this will be different from working in self-contained classrooms. Others expect counselors to provide career guidance for students without recognizing that counselors need time and resources to do this job well. Still others operate as though a new building, a new computerized grading system, or a new organizational chart would cure everything.

We have nothing against team teaching, career guidance, new school buildings, and the like. However, we have seen a number of single innovations seemingly get lost or fade away because other changes were not also made. The examples are numerous, but the point is that hit-and-miss change often turns out to be no change at all.

The Rand Study uncovered a number of projects that were not implemented in the sense of reaching their objectives or having "staying power." Instead, the projects were "implemented" in *pro-forma* fashion, they broke down altogether, or they were co-opted by project participants so that the project was changed to fit traditional ways of doing things.

Projects in which participants did carry out plans for new action were characterized by *mutual adaptation*. This phrase applies to an implementation process in which the project was modified *and* in which the organizational relationships among staff and among teachers and students were altered.

The study did find that some school-communities and districts were more receptive than others to innovations but, more important, found that the receptive setting was *not* a sufficient condition for effectively implementing new ways of acting. An implementation strategy that promoted mutual adaptation was critical.

Four strategies seemed most often to promote mutual adaptation or to allow it to occur: (1) on-line planning efforts and continuing reassessment of the ways the project was moving, (2) provision of extensive training for all participants, (3) frequent regular meetings of project personnel to discuss problems and to share ideas, and (4) the local development of materials. "Learning by doing"—even if it meant "reinventing the wheel"—helped particpants to identify with project goals and precepts.

Thus, purposeful and conscious change seem to be insufficient by themselves. For a change to have staying power, it must take into account all of the interdependencies among personal behaviors, norms and structures of their organizations, availability of materials, and the limitations or potential of the physical plant. At the same time, the change must be focused and carefully monitored to insure that it does not limit the availability of human resources for future change when that is called for.

Imposed Change

An increasing amount of evidence suggests that change interventions initiated or imposed from the "outside" are likely to fail. The Rand Study, for example, found that most implemented projects were initially based on information or treatments that were already known to local participants. Few projects were based on information gathered through a systematic "search for alternatives." Local actors—in a single school community or district—were likely to be skeptical about the reported "success" of some method that had been tried "somewhere else." They tended to rely on the advice of those they believed to have a thorough knowledge of particular and peculiar conditions in their school community or area.

However, outside agencies such as state departments of education, textbook publishers, and institutions of higher learning continue to act as though the impetus for educational change has to be externally imposed. Certainly teachers, parents, and counselors confirm this assumption each time they cope with an imposed change instead of initiating their own. Instead of predicting needed changes and actively searching for solutions, insiders encourage imposition by fighting fires as they wait until a "solution" presents itself. But then, because the imposed solution rarely solves the important problems, the need for fire fighting maintains itself.

Imposed change is often like hit-and-miss change in some ways. If many of the impositions or the "hits" come down on the same place, there is the real possibility of innovation overload. We know of one small elementary school, for instance, that tried to implement team teaching, differentiated staffing, a new reading program, and minicourses all in the same year. The staff got so tired, the students and parents so confused, and the principal so discouraged that almost everything was abandoned the next year.

In many ways, therefore, it seems that projects designed by "outsiders" will generally fail to gather support or to achieve their objectives. "Outsidedness" can and often will be defined in very broad terms. Teachers in "other" schools, personnel from the district office, and consultants from state departments or universities will often be categorized similarly. They will lack credibility because they do not "belong" to the school community system where the project is being tried. On the other hand, those who clearly do belong—like counselors and school psychologists—have a much better chance to bring about important and lasting change.

SUMMARY

Chapter 1 has explored the need for change in schools, the necessity for clearly established goals, and the several types of change most frequently implemented by school communities, together with the degree of potential success possible with each.

NOTE

1. Other categories of perspectives described by Havelock include: (1) the research, development, and diffusion perspective that emphasizes rational sequences of goal setting, planning, implementation, and evaluation behaviors as performed by various institutions; (2) the social-interaction perspective that focuses on the information to be exchanged and who influences whom in the exchange process; and (3) the linkage perspective that emphasizes the dynamics among roles, institutions, and functions in the process of bringing about change in schools.

This chapter introduces three major ideas: (1) that schools are complex social systems comprised of functional subsystems; (2) that changing a system means changing certain processes in the culture of these subsystems; and (3) that certain forces influence the functions that facilitators of change can fulfill in a system. Four major principles are explained: (1) focus the change effort on changing functional subsystems in which interdependence is highest and with which other subsystems have recognizable interdependence; (2) change the norms and roles of communications, meetings, problem solving, conflict management, goal setting, and decision making, as well as the relationship among them; (3) know what others expect of you and be honest about what you would like to do; and (4) develop multiple bases of power from which to exert influence on the system.

Chapter 2

SOME PRINCIPLES OF SYSTEM CHANGE

In the first chapter we spelled out some assumptions about system change that we think are basic. But because assumptions alone do not provide a rationale for the strategies and guidelines presented in the next two chapters, we include this chapter to outline a few key principles. The chapter is divided into four major sections. The first describes a view of schools as complex social systems. The second outlines several interpersonal and group processes that provide targets for system change. Functions that facilitators fill in the change process are discussed in the third section, and the fourth describes variables that must be taken into account when deciding which function to fulfill.

Schools As Social Systems

When we say that schools are social systems, we simply mean that they are not mere collections of individuals who behave in disconnected or independently determined ways. In schools, people come together to perform specified tasks, the major purpose of which is to facilitate student learning. People in schools, like the components of any system, interact in fairly regularized and predictable ways. Their actions produce consequences that can be observed and are sometimes examined to guide future action. In spite of the fact that people in schools rarely view themselves as part of a social system, we think that this perspective is important. Particulary in large schools, it is difficult to picture a single, unified "system." One way to grasp the total picture is to look at *functional subsystems.* Begin by defining some goal that people strive to achieve.[1] The goal of facilitating student learning, for example, includes functions like teaching reading, teaching math, and providing student support services. Functions may be further dissected, as when "providing student support services" becomes counseling students about career opportunities, providing food for the lunch program, or assisting students who become ill in school. Once a function is defined, all the people who must perform well if it is to be fulfilled can be listed. For example, those who must function if students are to eat include those who order food, deliver the raw materials, prepare it, serve it, and release students from class at a specified time.

    Often, of course, the same people appear in more

than one functional subsystem. For instance, a counselor may talk individually to several students, attend a pupil personnel division meeting to consider scheduling, talk to a few teachers about problem students in their classes, and explain some aspect of the school's program to parents—all in the same day.

Although the individual's perspective of the system is largely structured by the face-to-face interactions he or she has as a member of several subsystems, important work is done and major goals are accomplished only through the coordinated efforts of sets of persons. Just as all individuals depend on each other to perform the subsystem's function, all subsystems are likewise *interdependent*. By virtue of bringing people together to achieve common tasks, organizations create subsystems in which every component or part affects every other component or part. What one person or group does is affected by and affects others.

In spite of the fact that people in schools are arranged in interdependent functional subsystems, few change strategies take this into account. Take, for example, the typical strategy of sending individual counselors to workshops or in-service classes. There, if everything goes well, a counselor may learn that method "X" is effective and fun to use. When he or she comes back to the school, wanting to try it out, others may disapprove, the needed materials may not be there, the schedule may be too inflexible, or something else about the subsystem of which he or she is a part may make the effort unnecessarily difficult. Perhaps even a better example is the case of the high school counseling department that wanted to involve teachers in making decisions about student personnel services, made that wish public, and then got frustrated because it didn't happen. The counseling subsystem clearly ignored the fact that many teacher subsystems depended on it—if not out of preference at least out of habit—to make these decisions for them. Since other subsystems did not change to recipro-cate the counseling subsystem's preferred function, no change occurred at all.

We wish to highlight here that this monograph is about changing systems and not about using groups or systems as a means to bring about individual change. To illustrate this difference, let us return to the previous example. To change a *system* so that influence over deci-sion making is more widely shared, one must intervene both with subsystems that have made decisions in the past and with subsystems that have not. This kind of sys-tem change does not happen when counselors go off to workshops and learn to plan collaboratively with teachers or when teachers take courses in counseling and guidance. We are not detracting from the worth of workshops and training programs such as those de-scribed by Jacobs and Spradlin (1974), but we are saying that system change requies a theory and technology un-like the concepts and strategies that promote individual growth.

The key principle here is that to change a system, one must change the subsystems in which interdepen-dence is highest and with which other subsystems have a recognizable interdependence. Attempts to change a sin-gle person *will not* have a significant or lasting impact un-less other people change too; if one subsystem is to change its functions, other subsystems must be willing and able to reciprocate.

SYSTEMS HAVE A CULTURE

Think about similarities and differences among schools that you have observed. You can probably think of numerous similarities—older people always teach younger people, spaces are always set aside for storage of equipment, or doors typically open and close at certain times. Yet the differences are striking as well. In one school, students are rewarded for the same behavior that is punished in another school. Teachers in one school expect the principal to do what teachers in another school expect to do for themselves. How people behave and how people are expected to behave are very different; the culture is different.

The culture or climate of a system is in many ways no more or less than the way people behave and are expected to behave. Lortie's (1975) definition of culture says that "culture includes the way members of a group think about social action; culture encompasses alternatives for resolving problems in collective life. The storehouse of ideas may exceed the variety of observable behavior within the group, for some possibilities may not find expression" (p. 216).

The culture of the school becomes visible when one knows the *norms* (behaviors that will be rewarded and punished) and *roles* (expectations that others have for the behavior of certain individuals). Change the norms and roles and you change the system. This is the second key principle.

The rest of this section illustrates what we mean by changeable norms and roles. We have categorized our examples under six headings (communication, meetings, problem solving, conflict management, goal setting, and decision making), not because we find six a magical number or because these particular six categories have been found to be most meaningful to the largest number of people. Instead, we chose them for convenience to parallel chapters in a book with which we have lived (Schmuck, Runkel, Saturen, Martell, and Derr, 1973). The categories let us include much of what commonly appears in books on group processes or organizational change in a way that can also provide structure to a subsequent chapter on tools and strategies for intervening. So now, asking the reader's acceptance of these categories, we will proceed with examples of how norms and role expectations can change.

Communication

Many norms and roles concern the subtle and complex process of communication. This process includes verbal and written messages as well as nonverbal tones, signals, and postures. It dictates who speaks to whom, when, and why.

Because of the complexity of communication, the chances that this process can go awry or need changing are particularly high. Communication may not occur at all, or miscommunication—particularly under stress or about topics in which people have a great deal of emotional investment—can be obvious. Often, the following unproductive norms exist in subsystems:

- Group members continually make their own position known and persuade others of its merits by talking past each other and refusing to hear what others say.
- Group members ignore each other's feelings, never checking on anyone else's internal state.
- Everyone is expected to talk loudly and emphatically, no matter what the issue.
- People are listened to as long as they talk about trivial issues, but people withdraw when certain topics are mentioned.

Changes in the communication process can, instead, produce norms like these:

- Group members continually check to make sure they understand what others are saying.
- Everyone is expected to show concern for the feelings of others by monitoring nonverbal cues and checking their impressions.
- Voices typically remain calm, but people are heard—even when they express strong emotions.
- Important and fundamental issues or topics are discussed publicly with those concerned.

In many ways the communication process is basic to all the rest. Helping people to express their opinions and desires clearly, developing norms that support directness, and teaching methods for listening with understanding are strategies that can bring about important system change.

Goal Setting

We have often laughed at the truths in a story by Mager (1972, pp. v-vi) about King Aling in the land of Fuzz. It seems that King Aling commanded his round and multilegged cousin Ding to "Go into Fuzzland and find the goodest of men, whom I shall reward for his goodness." Ding impertinently asked how he would know the goodest of men when he saw one. The king whacked off Ding's leg for this affront and said only that the goodest of men would be "sincere." Ding limped out to search, but returned discouraged to ask his impertinent question again. Replying that the goodest of men would be "dedicated," the King whacked off another of Ding's legs. The story goes on for several more rounds until the King whacks off Ding's last leg and he falls to the floor with a "squishy thump." The "moral," says Mager simply, is "if you can't tell one when you see one, you may wind up without a leg to stand on."

Unfortunately, lots of educational goals are just as ambiguous as King Aling's. The following system norms apply:

- People take goals for granted, never discussing them or writing them down.
- Because no one can that say you do not get some place if you do not say where you are going, people are rewarded for ignoring goals.
- No one expects that progress toward goals will be carefully assessed.
- People state very different versions of goals that are supposedly shared.

Strategies that help people to sharpen goal statements or uncover and share goals that are not explicit can create norms like the following:

- No one discusses goals without ascertaining how much others are committed to them.
- People expect goals to be precisely stated, even when being clear makes it harder to avoid talking about conflicts.
- Many kinds of informal and formal procedures are used to assess movement toward goals.
- People are rewarded for getting where they said they wanted to be.

Problem Solving

The ultimate test of an effective system is its ability to solve the problems it encounters. For effective problem solving to occur, group members must identify discrepancies between the way things are and a more desirable condition. Once the problem is defined, group members must identify quality solutions that move them to where they want to be. An ineffective problem-solving process might include the following norms:

- Group members ignore problems or state them in forms that do not address the real issues.
- People expect others to solve their problems, rarely forming groups to do so.
- Problem-solving procedures are not systematic or procedures are rigidly followed without review or consideration.
- Since life seems happier when there are no problems, people are rewarded for withholding problematic information.

It is possible for groups to develop norms that support an effective problem-solving process. Examples of this kind of norm include the following:

- Group members state problems precisely and directly, accepting any discomfort as temporary and solvable.
- People organize quickly into groups to inquire jointly into common problems.
- Everyone understands and is expected to follow flexible, agreed on procedures.
- People are rewarded for viewing problems as normal and for viewing logical problem solving as a springboard to creative action.

Many systems would change markedly if people clearly defined the problems confronting them and developed workable plans in collaborative ways. Strategies that aim toward this kind of goal can make important system changes.

Meetings

If the most abominable thing about the system in which
you work is the process people go through in meetings,
this may become your first target for system change. For
example, you may very much want to change norms like
these:

- People straggle in and leave early, but no one
  pays enough attention to catch them up or fill
  them in later.
- Only certain people are expected to convene or
  take minutes.
- Everyone talks at once or some people do all the
  talking while others remain silent.
- Everything is "all business"; no one pays attention to others as people or to the way in which
  the work is accomplished.

By contrast, some groups have meeting norms that
we think are much easier to live with. It may take a con-
certed effort to turn things around, but consider the advantages of norms like these:

- Expectations about starting and ending times
  are clear and people help those who can not
  meet them.
- Convening and recording functions are shared
  or rotated so that everyone has a chance to participate without this added responsibility.
- Procedures encourage participation by all those
  who want to speak because monologues are not
  reciprocated by silence.
- Certain times are set aside to discuss the way the
  meeting went and any interpersonal feelings it
  created.

Decision Making

Creative efforts to solve problems include choosing among alternative goals, procedures, resources, and solutions. Even when only one solution is thought of or is possible, problem solving always ends with the choice of acting or not acting. The decision-making process in a system is activated any time members carve out a choice for action. This process is important because it often produces much conflict in the system. Norms like these often inhibit the way people exert or accept influence and make their decisions:

- Only those with legitimate authority are expected to make decisions.
- Responsibilities for decision making are vague or unclear.
- Group members are locked into a single method of group decision making—majority voting, for example.
- Even decisions that require everyone's understanding and commitment are made by one person or majority vote.

Through effective intervention strategies, it is possible to change system norms into something like the following:

- People with information make decisions because decision quality is viewed as more important than who makes it.
- Everyone has clear expectations about who will make what decisions.
- Procedures for making a decision are carefully matched to the kind of decision being made.

- Those who must understand and be committed to the decision are rewarded for participating in it.

Conflict Management

The process of identifying and managing or resolving conflict in a group is often quite easy to see. Consider a group with norms like the following:

- Conflicts are hidden from public view entirely or are only dealt with behind the scenes.
- People are rewarded for ignoring little conflicts and dealing only with those that have reached a sizable magnitude.
- Procedures are based on the expectation that conflicts are resolved only if one party loses while the other wins.
- People confuse disagreements with conflicts that arise from differences in self-interests and collaborative compromising with negotiation.

Norms like these are not only common in groups, they also pose a major hindrance to effective group functioning. Since conflict does not surface until differences of opinion or self-interest are expressed, and since different opinions can be a group's most valuable resource for creative problem solving, ignorance or suppression of conflict can stop a change effort cold. We think it is much more productive to change norms governing conflict so that the following behaviors are expected, exhibited, and rewarded:

- People admit they have different views and consider the tension as a signal to problem solve.

- Small conflict situations are handled as they occur.
- Others actively try to help persons engaged in
  conflict by serving as interested and caring third
  parties.
- People attempt to resolve differences of opinion
  in a collaborative, win-win fashion while skillfully
  negotiating differences in self-interest.

The process of conflict identification and the processes of decision making, meetings, and the like are
highly interdependent. For example, norms about who
may talk about what may enhance or limit the amount of
information that is available for finding solutions to
problems. Or, as another example, if few people are expected to take roles in the decision-making process,
there is high likelihood that self-interest conflicts will
surface.

Furthermore, the processes can be changed by persons acting from a variety of stances or roles. In the next
sections, we explain six roles that can be assumed for the
purpose of facilitating change in system processes. We
also explain some factors that should be taken into account when determining which role to assume.

Functions Filled By Facilitators of Change

We contend that the person who decides to bring about
change in his or her system can do so from a variety of
vantage points. We have identified six vantage points
in this monograph: member, convener, resource linker,
process consultant, third party consultant, and teacher
or trainer. The choice among these roles or vantage
points is never exactly clear, since the roles have at least
as many similarities as they have differences. Just as parents do a lot of teaching and teachers do a lot of par-
enting, a group member may perform the resource

linking function and a person designated as trainer may perform the convening function.

Being an Effective Group Member

Everyone who reads this paper is undoubtedly a member of several task groups or subsystems. As a member a person brings about system change primarily by modeling and risk taking. For example, the counselor in the pupil personnel department of a large high school might improve the effectiveness of this subsystem by continuously clarifying what others are saying. If others rarely make clarification statements and then begin to copy this behavior, the modeling would have its desired effect. Risk taking might be displayed by stating one's feelings about membership in the group when norms support staying on the task. Many of the examples in Chapter III can be translated into behaviors for modeling that require greater or less risk taking, depending on present norms.

Serving as Group Convener

Often groups can be helped a great deal by those who agree to serve as chairpersons or conveners. In Chapter III we suggest a number of tools that conveners can use in tasks such as preparing for the meeting, helping the group to warm up, helping group members to organize their agenda, encouraging wide participation, helping with record keeping, and leading the group in assessment of its own functioning.

Resource Linking

Sometimes groups and susbystems within an organization are fairly clear about existing problems, but lack the skill or resources for solving them. Valuable assistance can be provided in this situation by clearly offering one's own ideas and energy or by eliciting information about the resources that others have to offer. When the group needs resources that none of its members can provide, the resource linker searches for and suggests other people, materials, programs, or the like for examination. Functioning as a resource linker, the school psychologist might do demonstration teaching, set up a cross-age helping program, organize volunteer parents, or track down a curriculum consultant for teachers in a particular department.

Providing Process Consultation

Too often groups are unaware of how they are influenced by norms and group processes. Members of the student council may know that their meetings do not proceed smoothly, but they may not understand why. Teachers in a certain department may ask why decisions that they think they make never get carried out. Many of the tools presented in Chapter III can be used to highlight important group processes as the process consultant collects and feeds back data about actual and preferred agreements or norms.

Third Party Consulting

The functions of a third party consultant are not all that different from a process consultant. We have isolated this role only because it is most often played when problems occur with the process of uncovering and managing conflict. Third party consulting requires special skill in helping parties to a conflict to clarify their positions, establish working agreements, and use their differences as a springboard to problem solving. More about this function will appear in the next chapter.

Teaching Alternative Modes

Once in a while persons in a system may be asked by others to design and carry out a training intervention for the group. For example, a counselor might teach others on a district task force to make decisions by consensus if majority vote procedures seem unsatisfactory. Or, as another example, the school psychologist might help a teacher to train students in basic interpersonal communication skills. Some of the lessons in Chapter III and some of the guidelines in Chapter IV are directly applicable to the functions of training or teaching alternative modes.

DETERMINING THE STANCE TO TAKE

Selecting which function to perform can depend on many factors. These factors include very obvious ones such as what one wants to accomplish and what one has the skills to do. Primarily, however, we think that the choice is most often limited by how definitely others expect the facilitator to advocate a particular change or to remain neutral with no personal investment in specific outcomes.

For example, a colleague of ours once worked in an organization that was faced with the difficult decision of narrowing its scope of work and staff size to save money. She knew this decision was a unique one and likely to produce a lot of conflict. Before deciding which function she could play to be most helpful, she encouraged staff members to discuss what they expected of each other at the decision-making retreat. Since most people expected everyone to participate as an active member, they decided to hire two consultants from outside the organization to serve as process consultants. Her functions as member and resource linker became clear. Only a few months later, one of the subsystems of the organization wanted to spend a day on problem solving about roles and commitments. Since she was not a member of this subsystem, she was called on to serve as process and third party consultant. Her other skills could now be used because others expected her to remain sufficiently neutral about this problem.

Insidedness and Outsidedness

In the jargon of interventionists, expectations like these determine one's "insidedness" or "outsidedness." Insidedness refers to expectations that you will abide by the group's norms, recognize your interdependence with others, and work hard to get changes that will benefit your individual position. By contrast, outsidedness refers to expectations that you will remain neutral and objective at all times because the group can make certain changes that will not directly affect you. Either set of expectations can apply to any member of a group at certain times; you do not have to be a paid consultant to the group to be seen as outside a particular problem.

Insidedness or outsidedness *vis à vis* the system to be changed is accompanied by a different perspective. Remember the old adage about standing so close that you can not see the forest for the trees? That is the perspective of the insider. Before jumping to the conclusion that it is a bad place to be, consider how this gives one access to information to which no outsider is privy. If you really want to eliminate unproductive norms, install new norms, or otherwise change the system, some of that information may be vital.

But insidedness has its limitations. We have already mentioned that change takes energy. The stamina of change facilitators will be tested many times as they initiate change ideas and expend their personal resources to bring changes about. But unproductive norms also consume a lot of energy. It is possible that a system can drain people inside it to the point that they have little left to give.

The expectations of others thus may be a double-edged sword; being too inside or too outside can limit one's effectiveness. But the principle embedded in all this is simple: be clear about what others expect of you and honestly tell them what you would like to do. In this way, no matter which side of the sword you stand on, you can use it for a lever.

A Base of Influence

So far we have listed three key principles for changing a system: (1) change the functional subsystems in which interdependence is highest and with which other subsystems have recognizable interdependence; (2) change the norms of communication, meetings, problem solving, conflict management, goal setting, and decision making as well as the relationships among them; and (3) know what others expect of you and be honest about what you want to do. Before we conclude this chapter, we would like to offer one more principle: (4) develop multiple bases of power from which to exert influence on the system.

We predict that many people reading this monograph will begin to mutter as they read that principle. "After all," they'll say, "I'm not in any position to facilitate change legitimately," or, "The principal (or boss or supervisor) is the only one around here that can change anything." Before the muttering goes on too long, let us say that this view is partly right. There is a kind of power that comes from holding a position in which authority is legally vested.

But the view is also partly wrong. There are at least

four other bases of power, according to a model provided by French and Raven (1959). First, there is *reward power* that is based on the control of, and ability to distribute, reward valued by others. Even though someone else distributes paychecks or writes recommendations, others in a group are bound to have some resources— good ideas, a jovial manner, a comforting way—that are valued by others. This base is at least accessible to most people.

Second, there is *coercive power,* the opposite of reward power. This power is based on the ability to inflict punishment or withhold valued rewards. Few people deliberately try to develop this base of power because it sounds so cruel and inhumane; but remember that if people appreciate what you do, they are likely to resent you or fear that you can stop. Building reward power can have the side advantage of building coercive power as well.

Third, and most important in our minds, is the power that comes from knowledge, skill, or experience. We rate this as important because it is the easiest to develop. We hope this monograph contributes at least in some small way to the *expert power* base of readers.

The fourth and final base of influence is called *referent power.* French and Raven define it as based on personal attractiveness or membership in the primary reference group of others. Some may not want this kind of power and others may despair of ever having it, but it is important because it, too, can be developed with relative ease. Dr. David Berlew (as cited in Schmuck, Runkel, Saturen, Martell, and Derr, 1972, p. 12) described the components of charisma, a concept quite close to referent power. After studying great leaders to discover what motivated groups to follow them, he found three common characteristics: (1) the ability to relate goals to values cherished by the group, (2) the ability to make others feel stronger and possessed of a greater sense of personal efficacy, and (3) the ability to impart a sense of urgency concerning the stated goals. Convincing others—and maybe yourself—that you have this base of power to facilitate system change comes from learning and teaching others to say, "I want to do it!" "I can do it!" "Now!"

SUMMARY

In this chapter we have stressed that schools are complex social systems that can be changed only when their cultures are changed. In addition we have explained major principles that should guide the actions of those who would change the culture of a system.

NOTE

1. Both explicit and latent functions of schools need to be considered. By latent functions we mean things such as supplying jobs for adults, keeping children off the labor market, socializing students to prevailing values, etc. While this may seem cynical, we have observed that many people in schools expend a great deal of energy to realize goals such as these.

This chapter describes a number of tools and strategies used by facilitators of system change. We have sampled procedures, instruments, and activities for facilitators who are group members, conveners, resource linkers, group process and third party consultants, and skills trainers. Each highlights one or more group processes such as communication, goal setting, problem solving, meetings, conflict management, and decision making.

Chapter 3

STRATEGIES AND TOOLS FOR FACILITATING SYSTEM CHANGE

The predicament of many teachers has often been summarized as a tension between long-range planning and thoughts of "What will I do Monday morning?" We presume that a similar tension exists in readers of this monograph. Some, no doubt, prefer to read theoretical statements that help them to organize the tools already in their "bag of tricks." But some undoubtedly sense the near-emptiness of their "bag" and wish for tools and ideas that can be tried out tomorrow. For these people, the urge to have it all make sense will follow the urge to take action. We emphathize with this predicament. Sometimes abstractions and strategies are very seductive, but sooner or later come the hours when we sit down to design interventions that we will carry out.

This chapter is for those who wish examples of almost immediately useful ideas. It is a collage of questionnaires, exercises, checklists, procedures and observation instruments. Each highlights one or more of the group processes described in the previous chapter: communications, meetings, problem solving, conflict management, goal setting, and decision making. Some of these tools will be most useful to persons playing the process consultant role; other tools go more logically into the "bags" of skills trainers, third party consultants, conveners, or group members. Resource linkers may want to share these tools with others.

All of these tools and ideas have appeared before in many other sources. From time to time we will suggest references specific to an idea or book, but we have reserved most of our comments on other sources for the annotated resource section at the end of this monograph.

STRATEGIES TO IMPROVE COMMUNICATION

Have you ever seen a group where people did not talk about their reactions to each other? Where everyone carried a load of annoyances and hurt feelings? Where people were not sure that they understood or were understood? Quite clearly, groups like this have communication problems.

In other groups, members tell others how they come across and learn what their behavior does to others. "Gunny sacks" never fill up because they do not have to; information about personal feelings is shared regularly. People discuss their differences with the assurance that they have both the right to listen and the right to be heard.

Helping To Improve Interpersonal Communication

Four interpersonal communication skills are frequently

used in groups with positive communication processes. These include paraphrasing, behavior description, describing one's own feelings, and checking impressions.[1]

(1) PARAPHRASING.   Paraphrasing is any way of checking with others to be sure that you understand their ideas as they intended you to. Any means of revealing understanding constitutes a paraphrase. Paraphrasing is more than word-swapping or putting another person's ideas into other terms. Instead, it answers the question, "What does the other's statement mean to him or her?" and requests the other to verify the correctness of your interpretation. The other's statement may convey something specific, an example, or a more general idea to you, as in the following examples:

Other:	I'd sure like to own this book.
You	(being more specific): Does it have useful information in it?
Other:	I don't know about that, but the binding is beautiful.
Other:	This book is too hard to use.
You	(giving an example): Do you mean, for example, that it fails to cite research?
Other:	Yes, that's one example. It also lacks an adequate index.
Other:	Do you have a book on peer counseling?
You	(being more general): Do you just want information on that topic? I have some articles.
Other:	Great. Anything that I can read quickly will do.

(2) BEHAVIOR DESCRIPTION.   In a behavior description, one person reports specific observable behaviors of the other without evaluating them or making inferences ab-

out the other's motives, attitudes or personality. If you tell me that I am rude (a trait) or that I do not care about your opinion (my motivations) and if I am *not trying* to be rude and *do* care about your opinion, I do not understand what you are trying to communicate. If you point out that I have interrupted you several times in the last ten minutes, I would receive a clearer picture of what actions of mine were affecting you. Sometimes it is helpful to preface a behavior description with "I noticed that" or "I hear you say" to remind yourself that you are trying to describe specific actions. Consider the following examples:

> "Jim, you've talked more than others on this topic," instead of, "Jim, you always have to be the center of attention."
> "Bob, I really felt good when you complimented me on my presentation before the board,"
> instead of, "Bob, you sure go out of your way to say nice things to people."
> "Ellen, that's the fifth cigarette you've smoked in the past hour and the smoke is bothering my eyes,"
> instead of, "Ellen, you're deliberately polluting my air."

(3) DESCRIBING OWN FEELINGS.   Although people often take pains to make sure that others understand their ideas, only rarely do they describe how they are feeling. Instead, they act on their feelings, sending "messages" that others draw inferences from. If you think that others are failing to take your feelings into account, it is helpful to put those feelings into words. Instead of blushing and saying nothing, try "I feel embarrassed," or "I feel pleased." Instead of, "Shut up!" try, "I hurt too

much to hear any more," or, "I'm angry with you."

(4) CHECKING IMPRESSIONS. This skill complements describing your own feelings and involves checking your sense of what is going on inside the other person. You transform the other's expression of feelings (the blush, the silence, the tone of voice) into a *tentative* description of feelings and check it out for accuracy. An impression check (1) describes what you think the other's feelings may be; and (2) does not express disapproval or approval—it merely conveys, "This is how I understand your feelings. Am I accurate?" Examples include:

"I get the impression you are angry with me. Are you?"
"Am I right that you feel disappointed that nobody commented on your suggestion?"

Often an impression check can be coupled easily with a behavior description, as in these examples:

"Ellen, you've smoked five cigarettes so far and seem upset with the slowness of the meeting. Are you?"
"Jim, you've made that proposal a couple of times already. Are you feeling put down because we haven't accepted it?"

Many people learn and practice these skills in interpersonal communication workshops of various kinds and use them spontaneously in their interactions with others. As group members, counselors or school psychologists can model these skills. Such modeling may or may not be accompanied by direct explanations of the names and purposes of the skills. Third party consultants can provide a valuable service by explaining the skills and enforcing their use by the parties engaged in conflict.

One way a skills trainer can help people to learn the skills involves five steps: First, the trainer explains the skills and suggests a topic that people talk about (e.g., what I expect from this workshop or how I think parents and students could be involved in the school). Second, participants form trios and each person is assigned the role of sender, receiver, or observer. The sender begins the conversation and tries to describe his or her feelings or the receiver's behavior while discussing the topic. The receiver listens and either paraphrases or checks his or her impressions of the other's feelings. The observer notes instances of communication-skill use and reports his or her observations to the sender and receiver at the end of a specified time. Third, roles are exchanged so that a different trio member becomes the sender. A new topic may or may not be assigned by the trainer for this second round. Fourth, in the third and final round of the exercise, participants once again exchange roles. Finally, all trios may join the trainer in a discussion of how the exercise went and how the learning may be applied in day-to-day interactions with each other. In this debriefing session, the trainer should monitor and encourage use of the skills.

Modeling the skill of open communication, especially when the norms of the group support other kinds of communication behavior, involves taking risks. This is especially difficult when trust—the belief that others will not take unfair advantage and are motivated and competent to help you—is low. Openness, risk taking, and trust are paradoxically interdependent. Trust is established and maintained by openness, and openness is risky if trust is low; but only a small amount of trust can be built if only a small risk is taken.

Those who want to change communication norms

have to take this risk; there is really no other choice. It may be easier to do after practicing with a friend,[2] by concentrating on one skill at a time, or when you tell others in the group that you want to try modeling these skills because you think they are helpful.

Helping Groups To Perform Necessary Communication Functions

Task groups with effective communication processes do more than exchange interpersonal feedback. They have an agenda for task meetings and can also exchange information on procedures to use as they work together. They following observation tool can be used to check how often group members perform the task and group maintenance functions required in an effective group discussion. To use it,[3] simply place a tally each time a person performs one of the functions.

Task Maintenance Functions

Task Functions
1. Initiating: Proposing tasks or goals, defining a group problem, suggesting a procedure for solving a problem, suggesting other ideas for consideration.
2. Information or opinion-seeking: Requesting facts on the problem, seeking relevant information, asking for suggestions and ideas.
3. Information or opinion-giving: Offering facts, providing relevant information, stating a belief, giving suggestions or ideas.
4. Clarifying or elaborating: Interpreting or reflecting ideas or suggestions, clearing up confusion, indicating alternatives and issues before the group, giving examples.
5. Summarizing: Pulling related ideas together, restating suggestions after the group has discussed them.
6. Consensus-testing: Sending up "trial balloons" to see if

the group is nearing a conclusion, or agreement has been reached.

Maintenance Functions
1. Encouraging: Being friendly, warm, and responsive to others; accepting others and their contributions: listening; showing regard for others by giving them opportunity or recognition.
2. Expressing group feelings: Sensing feeling, mood, relationships within the group; sharing one's own feelings with other members.
3. Harmonizing: Attempting to reconcile disagreements, reducing tension through "pouring oil on troubled waters," getting people to explore their differences.
4. Compromising: Offering to compromise one's own position, ideas, or status; admitting error; disciplining oneself to help maintain the group.
5. Gate-keeping: Seeing that others have a chance to speak; keeping the discussion a group discussion rather than a one-, two-, or three-way conversation.
6. Setting standards: Expressing standards that will help group to achieve, applying standards in evaluating group functioning and production.

This observation schedule might be used by a convener during a meeting. Although it would be a full-time job to think about all the categories simultaneously throughout the meeting, the convener could work from one task category and one maintenance category for each agenda item. The convener might make a special effort to perform the specific functions he or she is attending to or might encourage participation of those group members who typically perform these functions well in group meetings.

Observations such as these could be summarized and reported back to the group by a process consultant to start a discussion of how the group perceives the nature and intensity of its communication problems. It would also be possible to use this list of functions as the basis for an exercise in which each person in the group would write down the names of persons who consistently and effectively perform each function. Individual nominations could then be tallied on a large chart and the group could discuss how task and group maintenance functions were dispersed among the membership. It would be important here for the process consultant to explain that everyone in the group—not just the formally designated leaders—can and often does take responsibility for performing these functions.

If the process consultant is worried that posting the names of individuals on the chart will cause too much discomfort for some group members, he or she might tally the observations at another time and record the nominations for Person 1, Person 2, etc. Although the feedback should be presented in a way that captures the interest of group members, efforts to insure anonymity of nominators and nominees may prevent raising unnecessary anxiety that could prevent exploration of the issues.

Helping Groups To Meet Identity Needs

These issues evidence themselves as group members try to answer questions like: "What's my role in this group?" "Who's in and who's out?" "Should I be assertive or quiet?" These questions are often particularly common in groups that are newly formed, have new members, are large, or meet infrequently.

Conveners or process consultants can help group members to understand clearly what they expect of each other. For example, if a new member has been asked to join the group to represent the perspective of a particular constituency, that expectation ought to be clear. Or, if the faculty has created an ad hoc task force to accomplish a given task by a given date, the nature and timelines of the task should be specified.

Process consultants or trainers can also recommend "warm-up" exercises in which people get acquainted with the roles each wants to play. We have frequently used some version of the "Who am I" exercise (see Johnson, 1972 for details of several versions). In one version, each person writes five or ten answers to the question, pins them on his or her chest, and mills around to read others' answers. In another version, people answer the question with phrases and pictures on a large piece of newsprint, posting their murals on the wall for all to study. Many other exercises in Johnson (1972) and in Simon, et al. (1972) are also useful as warm-ups.

Helping Groups To Resolve Control Issues

All people have some need to feel influential and powerful, and the way this need manifests itself in a group may take many different forms. We have attended numerous faculty meetings, for example, in which competitive remarks, passivity, or meaningless arguments suggested to us that issues of who will control or defer to whom were salient. Wise conveners and process consultants recognize that it is more efficient in the long run to get the control issues out from under the table and to make them legitimate topics of discussion.

Control issues often surface and can be dealt with as the agenda for the meeting is established. If the convener prepares the agenda in advance without having others participate, this opportunity is lost. The following procedure,[4] by contrast, assures that all participate in agenda building and have the chance to explore how they can and will influence the rest of the meeting. The procedure has four steps that a process consultant, convener, or group member can introduce.

1. The convener asks everyone to brainstorm[5] possible items for the agenda. These are recorded on newsprint

or a blackboard so all can see.

2. Items are discussed until everyone is clear about which are information items, which will involve making agreements or decisions, and which call for sharing or discussion.

3. The convener asks for suggested sequences, making sure that priorities are discussed. Numbers to indicate the agreed upon sequence and names of persons presenting each item are recorded.

4. All members debrief the agenda-building experience by describing (a) their own feelings about the procedures, particularly their feelings about low or high influence over the agenda; (b) how others behaved to influence the agenda; or (c) their perceptions of how influential others felt.

Control issues can also be dealt with during the meeting. Group members or process consultants can listen carefully for proposals (suggestions of what to do or not to do) and note what happens to them. The following possibilities[6] exist:

The Plop: No one responds to the proposal in any way.

Self-Authorization: The proposer implements the decision without explicit agreement from anyone.

Handshaking: One other person supports the proposal.

The Veto: One other person negates the proposal. Often support or criticism of the proposer accompanies the handshake or veto.

Majority Rules: More than half (or some other pre-established proportion) explicitly agree to implement the proposal. The intentions of the minority are not clear.

Consensus: Everyone evidences understanding of the proposal, describes his or her feelings and opinions, and either (a) supports the proposal because he or she thinks it definitely will work, or (b) says publicly that he or she will give the proposal an experimental try even though he or she has doubts or would have preferred an alternative.

Unanimity: All group members explicitly approve the proposal.

We have observed that many groups strive for unanimity on every issue and are disappointed when it cannot be reached. As process consultants we have found it useful to explain that unanimity assumes a fixed solution, as when a jury must make a unanimous decision that the defendant is either innocent *or* guilty. Unanimity is thus very hard to achieve for many kinds of issues and requires much "people-shaping" to change the attitudes of those in the minority. Consensual decision making, on the other hand, is best seen as "solution-shaping."[7] Group members who object to the proposal can be asked how the solution would have to be changed to meet their needs, and no one expends energy trying to shape or change the needs, attitudes, or values of objections.

Since consensual decision making is frequently the least typical mode of decision making in educational groups, a skills trainer might want to suggest exercises through which groups can practice consensual decision making and become aware of its benefits for decision quality and acceptability.

Some exercises that can be used for this purpose are described in Schmuck, Runkel, Saturen, Martell, and Derr (1972, pp. 273-280) and also in a later section of this chapter.

Helping Groups To Reconcile Individual Needs and Group Goals

According to Schein (1969), member needs must be exposed and shared to some degree before it is possible to set up valid group goals. Therefore, every meeting should include some activities in which members get a chance to express what they want to get out of belonging to the group. The agenda-building procedure described in the previous section can accomplish this, as can a procedure called process debriefing. Process debriefing usually occurs at the end of the meeting or at the end of each agenda item if these are particularly long. Each member shares his or her impressions of how the group is working together and can state if the processes being used are meeting his or her needs.

Examples of possible statements during a process debriefing session include:

1. "I've noticed that three or four people have been doing most of the talking. I haven't participated as much as I want to and feel ignored whenever I start to speak."

2. "I'm happy that we've stayed right on the topic. I wanted more than anything to get on to item five, and it looks like we'll make it."

3. "Jim, I really liked it when you encouraged Bill to tell you what he thought about your behavior. I want us all to be candid and receptive with each other."

A possible variation to debriefing is to have each member write down before the meeting starts how he or she wants to feel at the end of the meeting. Then each takes a few minutes at the end of the meeting to share his or her present feelings and to describe how they compare to the initial wants. Debriefing proceeds as the group makes agreements about how the next meeting should go so that more people can have more of their wants met.

However, process debriefing may break group norms or taboos against being open, discussing feelings and behavior; suggesting that the agenda include time for this may not always be easy to do, especially by group members. One "trick" that we have seen group members use to alter the taboo is first to make the suggestion for debriefing when they think someone *else* in the group is feeling bad and would like an audience. For example, a group member might say, "I've observed that several people in our group have been interrupted a lot in the meeting. If I had been in their shoes, I'd like to talk about how I felt put down. I felt pretty good about the way my ideas were received today and would like us to offer help to those who didn't. I'd like to learn what I could do to help our meetings so everyone feels okay when they are over." The basic rationale behind this strategy is that asking for "air time" to *offer* help is often more successful than asking for "air time" to describe one's own grievances or to ask for help. It may take repeated offers, but others may eventually come to value the offerer's learning goals enough to participate in debriefing.

Helping Groups To Stimulate Acceptance and Intimacy

Every group develops norms that support a certain range of accepting and intimate behaviors. In some groups where informality reigns, people call each other by their first names, personal exchanges are desirable, and procedures are not written down. In other groups, people are extremely polite with each other and written procedural models guide what is said and done.

There certainly is no level of acceptance and intimacy that is "best" for all groups at all times; groups may need more or less stimulation to meet member needs for interpersonal closeness. Conveners and process consultants can be aware of this issue and help the group recognize that it is legitimate to do problem solving about these norms.

A questionnare[8] to elicit information about how members perceive group norms of interpersonal support follows. Each member makes two marks on each line, an "X" to indicate what group members typically do, and a

"Y" to indicate how they would like group members to behave.

Group members evaluate or judge each other's behavior	//////	Group members describe each other's behavior
Group members attempt to control or change each other	//////	Group members collaborate in solving problems of mutual concern
Group members use "strategies" and preplanned techniques on each other	//////	Group members are spontaneous, straightforward, and honest with each other
Group members seem neutral, detached toward each other	//////	Group members convey empathy and respect for each other
Group members make statements about their superiority to others	//////	Group members attach little importance to differences in ability, worth, status, or the like
Group members express certainty and dogmatism, defend their ideas as "truths"	//////	Group members communicate provisionalism, a willingness to experiment or remain open to new ideas

Individual ratings can be tallied, and differences between the means of "X's" and "Y's" can be compared to start the discussion. Group members should be encouraged to use the communication skills during the discussion. It might help to have them write examples of behaviors or their own feelings to help explain their ratings.

STRATEGIES FOR IMPROVING PROBLEM SOLVING

People who would change a system must solve many kinds of problems. There are at least two different sets: (1) problems of how individuals may contribute to changing the system, and (2) problems that the system has to solve in order to accomplish its mission and to maintain itself as a system. Since most of this monograph examines ways to solve the first set of problems, let us concentrate here on the nature of system problem solving. Some of the tools presented may, of course, also be used by individuals as they consider how to help others.

As is true of the number of educational problems that need to be solved, there are many approaches to problem solving. Some include a linear sequence of phases: (1) identify the problem or assess needs, (2) agree on objectives, (3) search for alternatives, (4) choose a means to reach the objectives, (5) implement the chosen plan, and (6) evaluate what happens. Other approaches emphasize the "closed loop" nature of problem solving with which a group may start at any step and recycle as needed. Some problem-solving technologies are designed to answer the question, "What is the best way to reach our goal?" Other technologies are more often used to find answers about "How and why are things happening as they are?" or "What shall we do given that different parties need limited resources to move toward different goals?"[9]

An extremely simple and very powerful model[10] that we like to use distinguishes three kinds of information and three kinds of problems with which groups deal

1. *Situational Information:* that which describes the current situation or condition.
2. *Target Information:* that which describes a preferred state of affairs.
3. *Proposal Information:* that which describes ways to move from the current situation to one or more targets.

A "problem" exists anytime there is a discrepancy between a situation and a target. A problem is identified any time people say, "The way things are is not the way we want them to be."

It is possible to start with any of the three kinds of information and elicit the other two. For example, if someone says, "Let's change the class schedule from a six-period day to an eight-period day" (a proposal), information about the situation and target can be elicited by asking, (1) "What about the six-period day makes it unacceptable?" and (2) "What are the desirable features of an eight-period day?" As another example, if someone says, "Too many students come in for counseling in the morning and too few come in the afternoon," a paraphrase like, "Would your goal be to even out the load on counselors during the day?" can elicit a target statement. Once the situation and target were clear, it would be possible to brainstorm ways to remedy the situation.

According to this model, the three kinds of problems with which groups deal are task problems, methods problems, and process problems. Task problems concern a gap between the goals and output of an organization. Examples might include the lack of a career education program in the school, the lack of procedures for identifying students who want to receive counseling, or the lack of close working relationships between the school psychologist and some teachers. A task problem can have *any* content, and it is usually recognized only by the fact that it is on the agenda. Methods problems, by contrast, do not get on the agenda but arise only as task problems are worked on. Methods problems might include the lack of an agreed on time to end a meeting, the lack of agreements about whether a vote will be binding, or the lack of clarity about what the recorder will write down for the minutes. Process problems arise as members use their methods to work on their task. They include all kinds of interpersonal issues such as the lack of chances to speak, a lack of trust among people, or the lack of warmth and intimacy someone wants.

This model can be used in a number of ways. A group member or convener might keep track of the frequency and duration of task, methods, and process discussions during a meeting. Since groups that do not discuss methods and process issues with some frequency often bog down in discussing task issues, the member or convener may want to ask questions like, "How are we going to decide?" "When will we have time to discuss this further?" "How does that agreement make you feel?" "Is everyone satisfied that his or her views have been heard?" Questions like these can make the group conscious of the need for all three kinds of problem solving.[11]

A process consultant might stimulate problem solving about the methods used in a particular meeting with questions like these:

1. *Eliciting Situational Information:* How do people feel about the way this meeting went? What are

the consequences of a meeting that goes like this one?

2. *Eliciting Target Statements:* How would you like to have meetings go? What should the group do differently next time?

3. *Facilitating Proposal Brainstorming:* What steps can the group take to make the next meeting more productive and satisfying? Who will do what?

A person in a resource-linker role pays special attention to *proposals*. Imagine, for example, a counselor who is helping a teaching team discuss the lack of mutual support between older and younger students. The counselor might propose a program of peer tutoring such as that designed by Lippitt, et al. (1971). If the team already knows of enough options and just cannot decide which to use, the counselor might propose that the decision be made by consensus, since everyone will help to implement whichever program is chosen. Finally, if debriefing seems to focus on unequal participation, the counselor might propose that one of the "low-talkers" serve as convener next time. The possibilities are endless,[12] and the resource linker's imagination for creative proposals or ability to encourage others to share their proposal ideas is basic.

STRATEGIES TO IMPROVE CONFLICT MANAGEMENT

Like death, taxes, and rain in Oregon, conflict in groups and organizations is inevitable. The inevitability and normality of conflict make it a vital topic for those who would facilitate system change.

Schmuck, Runkel, Saturen, Martell, and Derr (1972) describe its pervasiveness well.

> "It occurs within persons when they feel ambivalent or confused about a course of action or when their expecta-

tions are unfulfilled. It occurs between persons when their goals are frustrated by the other, when they are competing to try to win some reward at the other's expense, when they misunderstand or disagree with another's expectations of their conduct, or when they approach a problem from a different point of view. It occurs between working groups in the organization for many of these same reasons. It occurs between organizations when they compete for scarce resources or when, in the case of revolutionary movements, the legitimacy itself of some organization is challenged" (p. 136).

Furthermore, conflict in work groups is virtually unavoidable; it arises out of the interdependence that distinguishes groups or organizations from mere collections of individuals. Since individuals in systems can perform their jobs well only if others do likewise, the chances of failure are increased as much as are the chances of success.

Finally, we think it is important to say that conflict is *not bad*. A basic premise behind working with conflict is that it can be healthy. It creates the stress that makes groups produce more and change. Only the most stagnant, bureaucratic, and dull organization imaginable would be without it.

Deciding When To Help

Certainly if conflict is pervasive and if some of it can or must remain, a primary decision of the process or third party consultant is when to intervene. We recommend three guidelines. First, it is easier to work with conflicts when all parties to the disagreement can and will assemble. There is little point in trying to manage disagreements between people who are too angry to speak to each other. Ultimately they must resolve their differences through a compromise or trade-off unless one or the other party is willing to exit permanently. We have no suggestions here for dealing with students who want to stay outside and riot or with negotiators who will not come to the bargaining table.

Second, be sure that the conflict is serious and destructive instead of one that creates healthy tension. It is quite easy to ask the following questions:[13]

1. Which of the following best describes the seriousness of the dispute?
   ———not at all serious
   ———not serious
   ———average
   ———serious
   ———very serious
2. Which of the following best describes consequences of the dispute?
   ———makes me (us) work much better
   ———offers encouraging competition
   ———doesn't affect me (us)
   ———is keeping me (us) from being as effective as I (we) could be
   ———is destroying our working relationships

Third, when given the choice of two conflicts that are similar on the first two guidelines, work first on the one in which the parties are most interdependent. It would be more important, for example, to help two teachers who had to team-teach a class than to spend time with two teachers who shared few responsibilities.

Because interdependence exists in the perceptions and expectations of people, ascertaining interdependence is not as simple as checking where the lines are on an organizational chart. Answers to the following questions can help to define critical interdependence:

1. On whom do you depend most to do your job well?
2. If you want to achieve a pesrsonal goal, whom do you ask to work with you?

3. Were lightning to strike all but two people in the group besides yourself, which two others is it most important to spare?

After all the members of a group give their responses to one or more of these questions, a diagram connecting those who nominate each other can depict interactions where interdependence is highest.

Even more data can be obtained from a pair of items such as the following: (1) name three people on whom you depend most to do your job; and (2) for each person you listed, name three people on whom they depend most to do their job well. Again, sociometric-style diagrams can display the varying intensity of interdependence.[14]

Consulting as a Third Party

In a short and excellent book, Walton (1969) described an approach to what he calls "interpersonal peacemaking." His methods are particularly aimed at helping people to resolve conflicts over ways and means when differences in self-interest are *not* paramount. As we said earlier, conflicts that stem from differences in self-interest require negotiation, not collaborative peacemaking. Although it is not possible here to give the topic of peacemaking all the attention it undoubtedly deserves, a quick summary of some key points may give the reader a rough idea of the principles Walton recommends.

First, the third party consultant tries to equalize the motivation of both parties to reduce the conflict. The more highly motivated person can be encouraged to slow down just as the person with low expectations about outcomes can be encouraged to hope for more.

Second, the consultant tries to equalize the situational power of both parties. Allies for the person with less power can be sought, and a rule about taking turns can help the person with less verbal ability. Even the seating arrangement or place in which the confrontation occurs can be chosen to equalize power.

Third, the consultant can help both parties to achieve a similar definition of the situation. Through preliminary interviews, the third party can ascertain whether one person's desire to confront is reciprocated by the other's receptivity.

Fourth, Walton maintains that the potential for resolving a conflict—achieving integration, in his terms—is no greater than the adequacy with which the parties have defined the differences between them. In other words, an effective intervention requires a stage in which the two parties are encouraged to explain their positions before a stage in which they generate proposals to deal with their differences.

Fifth, the third party should provide a source of emotional support to those engaged in conflict. It may be necessary to work with others in the group to which the parties belong to encourage norms in support of dealing with conflict.

Sixth, the most important task of the third party consultant is to increase the accuracy of communication between the parties in conflict. Walton suggests "translating or restating the messages until the sender and receiver agree on the meaning; procedural devices which require one to demonstrate that he understands what the other has said; and contributing to the development of a common language with respect to substantive issues, emotional issues, and the dialogue process itself" (1969, p. 147).

Finally, the third party can try to increase or decrease the sense or urgency so that an optimum level of tension exists. This can be partly accomplished by encouraging both parties to meet now or by recommending that the confrontation be delayed for a specified time.

The following five-step procedure[15] gives an example of how a third party consultant can help two people who want to improve the way they work together.

Step 1. Each person makes up three lists:
(a) Positive feedback list: things the person values in the way the two people have worked together.
(b) "Bug" list: things the person hasn't liked or can't abide.
(c) Empathy list: a prediction of what the other has on his or her positive feedback and "bug" lists.

Step 2. Each presents his or her positive feedback and "bug" lists to the other; then they share their empathy lists. During this period, the third party discourages any talk not directed specifically toward gaining an understanding of the other's point of view.

Step 3. Each then offers any information which may clarify matters. Again, general discussion is barred.

Step 4. The parties now negotiate around changes they want. They consent to planned changes and then decide how they will work together to bring them about. The third party lists the agreed-upon actions to be taken. He or she also lists those issues still unresolved. The pair then decides how these will be dealt with, or perhaps agrees that they will be left unresolved for the time being.

Step 5. The third party and the pair plan followup measures. Fordyce and Weil offer the following "operating hints" for the procedures:

"Not for the routine issues that come along daily. The method is designed for major overhauls.

Both parties must somehow arrive at the meeting in a disposition of good faith, and both must repose confidence in the Third Party. A Third Party is perhaps even more important than in a team-building meeting where the larger groups provide a moderating influence.

Sometimes it is good to enlarge the meeting by including others who know the principals and who can offer valuable insight. This should be done only if both principals agree.

It is important to check the hearing mechanism of the two parties. A way to do this is periodically to ask one person what he thinks the other has just said.

The structure and discipline of the meeting can be shaped to the problems and the principals' skill in handling them.

Chart pads can be helpful, even for a twosome. Role reversal can be extremely effective in fostering empathy.

Remember that many interpersonal problems are created or relieved simply by new organizational arrangements and job definitions.

Follow-through is important. A followup meeting may be scheduled, or the Third Party may later touch base with the parties individually, and bring them together again if necessary" (pp. 115-116).

Strategies for Improving Goal-Setting

We have already made reference to some tools that can be used to define organizational goals. Agenda setting procedures listed earlier in this chapter can be used to set short-term goals, just as questions in the section entitled "Strategies for Improving Problem Solving" can elicit information about targets.

We include yet another reference to this process because goal setting is closely related to measuring a group or organization's effectiveness and because a basic way to compare two systems is in reference to each system's record in goal achievement. It would, of course, be ridiculous to compare the effectiveness of organizations with extremely different goals. People do not care about the ability of teachers to make flyswatters any more than they care about the track record of the Ace Flyswatter Company in helping students learn to compute.

Although it can be extremely difficult, it can be worth attempting to compare the effectiveness of systems with similar goals. Many school patrons are interested in the quality of education provided by different schools, just as many car buyers examine the quality of output from various motor companies as they shop.

Goal setting is an important process within an organization because the process is basic to problem solving and change, and it often figures prominently in interpersonal communication and conflict. Within an organization, we are most interested in two features of goals: (1) the degree to which they are clear, and (2) the degree to which they are shared. These features are related in that it is hard for people to be clear about goals they do not share, and it is hard to share goals when clarity is missing.

Procedures, techniques, formulas, and even elaborate systems for goal setting have proliferated in recent years. Pressures for accountability that spawned "management by objectives" also led to the creation of planning-programming-budgeting systems in education. Most states now have mandated that schools install some system to report how tax dollars are spent and what those tax dollars buy. All such systems, of whichever type, require use of the goal-setting process.

Clarifying Goals

Educational goals are notoriously unclear. How often, for example, have we heard that a school district intends to "prepare students to live in a democracy"? Several factors may account for this. First, goals may not be stated frequently enough. A picture of where people are headed can change each step of the way as plans are implemented and new problems arise.

Second, any individual statement of a goal should always be assumed to be tentative and incomplete. Multiple expressions of goals, sampled at different times, are more reliable.

Third, many educational goals—like the goals of all people-changing systems—seem to be inherently unclear. Recognition of this truism can help people to conserve their energy to get clarity and agreement where it is possible and important. For example, a group that focuses on "What should a counselor do with this kind of referral?" can lead to specific goal statements.

Clarity, however, cannot be obtained until goals are stated. One technique that can be used to elicit goal statements is, "Ten Years From Now,"[16] a procedure that fantasizes the future. Group members can describe what life *is* like ten years from now either by writing their ideas on paper or drawing a mural with colored chalk on newsprint. The descriptions can illustrate things such as (1) what I do every day, (2) what my responsibilities are, or (3) what outputs of our group I am most proud of. As all members of the group report their fantasies, others are encouraged to paraphrase.

Agreeing on Goals

All members of a group can be extremely clear about goals that they want to achieve and still have problems working together to reach them. Movement toward one goal may actually inhibit movement toward another, or differences in priorities can cause poorly coordinated action.

A procedure developed by Helmer (1966) can help even extremely large groups to agree on top priority goals. The procedure includes five phases.

Phase I: All participants write goals they think are appropriate for the group. These are combined and tabulated by the consultant into a single list of ten or so that are most frequently mentioned.

Phase II: All participants prioritize the goals, writing a "1" beside the one they think is most important, a "2" beside the goal that is next most important, and so on. The consultant tabulates the ratings of individuals and presents a report of the findings.

Phase III: All participants again prioritize the goal statements. Those who deviate from the majority are asked to state their reasons. The consultant tabulates the ratings again and includes statements of reasons in a report.

Phase IV: All participants prioritize the goals one more time and the final report is prepared.

Phase V: If a high degree of agreement has not yet been achieved, participants may agree to debate, to divide into subgroups that pursue different goals, or to try another solution they propose.

STRATEGIES TO IMPROVE DECISION MAKING

It may seem strange to some that we isolate the process of decision making in a separate section. It is, of course, inextricably intertwined with communication, problem solving, meetings, goal setting, and conflict management. We have done so primarily for two reasons. First, decision making is a very salient issue to many people. They may not know or care much about optional ways to set goals, but they are likely to complain loudly about decisions they do not like or in which they had no chance to participate.

Second, we isolate decision making because it is often done sloppily, unconsciously, or without the involvement of people who want and need to participate. Clearly our biases and values enter in here to a great extent, and we make no apologies for that. The tools we describe in the following sections have helped many groups to make higher-quality decisions that were acceptable to more people; we think that is a valuable target.

Determining Participation

The literature of social-psychological theory and research on decision making is extensive.[17] People have studied those who make decisions, how decisions are made, and how others are affected when decisions are made for them.

Many recent writers have advocated participative management styles that allow all members to influence decisions that are to be carried out by the group. They have found that participation increases the satisfaction of workers and reduces the likelihood of sabotage. They have also established that the "boss" who allows subordinates to make decisions does not lose influence; it is possible for some group members to become more influential without other members becoming less so.

We have often used a questionnaire[18] to determine perceptions of and preferences for influence patterns in groups. The questionnaire poses two kinds of questions:

1. In general, how much actual influence do you think each of the following groups or persons *has now* in

determining what innovations are attempted in your school?

2. How much influence do you think these groups or persons *ought to have* in determining innovations attempted in your school?

Each question is followed by a list of persons and groups that includes the school board, the principal, you as an individual, a small group of teachers, parents, students, and the PTA. For each person or group, the respondent checks, "no influence," "a little," "some," "considerable," or "a great deal."

Of course, it is possible to change the questions to elicit information about influence over curriculum policy, codes for student behavior, teachers salaries, and the like. It is also possible to revise the list of persons and groups to list positions that people in and around the school actually occupy.

To use a questionnaire like this requires a four-step procedure. First, the helper (the convener or process consultant, usually) creates a sample instrument and determines who will be interested in discussing the finding and who should be asked to respond.

Second, the helper administers the questionnaire, perhaps by a "mail-back" procedure or during a meeting. Procedures for collecting the data are usually determined by the number of people who will respond, but they should always be designed to protect the anonymity of individual respondents.

Third, the helper summarizes the data to show the numbers of respondents who attributed or wanted various amounts of influence for various persons and groups. Statistical analyses of significant differences may or may not be used.

Fourth, the helper presents a summary of the findings to the group and leads a discussion of how the findings differ from the ideal situation, steps the group could take to solve problems suggested by the findings, etc.

Teaching Alternate Modes

We have found that many groups are dissatisifed with how decisions are made. They may complain that too many decisions are made by one person or a minority and that majority votes also fail to utilize everyone's resources.

We have also found that few groups know how to use any other method; they want to use procedures that guarantee that everyone can make his or her thoughts and feelings known, but do not know how. Skills training that helps a group learn to make consensual decisions is most effective when everyone in the group is motivated to find a new decision-making mode.

Training exercises[19] do not focus on real-life content; choices in an exercise are simulated so the group can focus on processes used to make the decision without concerns over long-term effects. Training episodes, because they are artificial, must be followed by discussions of how learnings can be transferred to the real world.

Whether the school psychologist or counselor chooses to offer skills training depends greatly on the willingness of others to learn new ways. In most circumstances, we recommend against trying to train others in one's own group, particularly if the counselor or school psychologist who wants the group to have training should feel this way. If his or her own role in decision making has been unsatisfactory, it is easier to let someone else manage the training sessions so that he or she can be involved as one of the trainees. Also, the simulated content of exercises is more easily introduced by a person who has less affiliation with the real decisions of the group. In short, while your system may require training to adapt new modes of decision making, we recommend that you give serious consideration to the person who should do this.

Clarifying Decision Making

We have repeatedly recommended survey-feedback techniques in this chapter because we think that they are effective and efficient means for facilitating system change. In this section we describe another survey-feedback tool that clarifies how different people are to be involved in various decisions. As with other tools, we expect the reader to substitute words freely when something else seems more appropriate.

This sample instrument is a matrix[20] that lists positions in the school across the top and clusters of decisions down the side.

	Administrators	Teachers	School Psychologist and Counselors	Students
Determining Curriculum				
Ordering Supplies				
Scheduling Students				
Evaluating Teachers				

Individuals put one of the following letters in each cell to indicate the kind of influence appropriately exerted by each occupant of a position on each kind of decision.

I - must be informed of the decision
C- must be consulted and allowed to influence
P - must participate, has a vote
V - has veto power, must agree
A - has sole authority to make the decision

Individual responses are then compiled and discussed; and goal setting, problem solving, or conflict management procedures are used as needed.

SUMMARY

Chapter III has offered suggestions and strategies for improving communication including paraphrasing, behavior reporting, and describing feelings. Ways are explored in which meetings of all kinds can be strengthened to meet group members' needs. When group members hold divergent views, the convener has a variety of alternatives through which group goals can be clarified, such as agenda building, brainstorming, and reaching consensus. Strategies for improving problem solving are presented, and numerous examples from the literature are included. The use of a third party to act not only as a consultant but also as a "peacemaker" between conflicting parties is examined through specific intervention steps. Exercises and questionnaires to expedite group goal setting and power structure are included.

NOTES

1. We have adapted descriptions of these four skills from some originally developed by John Wallen for the interpersonal communication training system of the Improving Teaching Competencies Program in the Northwest Regional Educational Laboratory.

2. For many exercises that can be used with a partner or in a small group to improve interpersonal communication skills, see Pfeiffer and Jones (annual handbooks) or Johnson (1972).

3. This observation schedule appears in Schmuck, Runkel, Saturen, Martell and Derr (1972, pp. 287-288). They adapted it from categories of task and maintenance functions in groups first distinguished by Ken Benne and Paul Sheets (1948).

4. Adapted from Schmuck, Runkel, Saturen, Martell, and Derr (1972, p. 190).

5. Brainstorming, as we use it here, simply means naming ideas as rapidly as possible without evaluating them. No one says, "That's not a good idea," or, "I really want to talk about that," during the brainstorming time.

6. These categories were originally shared with us by Fred Fosmire of the Psychology Department at the University of Oregon. He has used them in extensive work as a consultant for many business, industrial, and educational organizations.

7. The "people-shaping" and "solution-shaping" distinction was first explained to us by Matt Miles, one of the reviewers of this monograph. He has found it an effective way to help people understand what consensus is like.

8. Items on this questionnaire come from categories of defensive and supportive climates that were first described by Jack Gibb (1961).

9. We are indebted to Charles Jung, former director of the Improving Teaching Competencies Program of the Northwest Regional Educational Laboratory and a reviewer of this monograph for suggesting the questions that illustrate differences among (1) technical or systems problem solving, (2) theoretical or action research problem solving, and (3) values-based conflict and negotiative problem solving.

10. This model was first taught to us by Fred Fosmire of the Psychology Department of the University of Oregon.

11. Research by Maier (1970) provides very convincing evidence that groups that ignore problem solving about procedures and interpersonal issues have trouble solving task problems as well. He has also pinpointed ways that conveners can stimulate effective problem solving.

12. A wonderful collection of the endless possibilities can be found in Koberg and Bagnall (1972).

13. Adapted from Schmuck, Runkel, Saturen, Martell, and Derr (1972, p. 156).

14. For a general discussion of ways to help people deal with conflict, especially in classrooms where interdependencies are sometimes very high, see Gordon (1974).

15. This procedure is adapted from one explained in Fordyce and Weil (1971, pp. 114-116).

16. This procedure is a modification of one called "Projection into 1982" from Schmuck, Runkel, Saturen, Martell, and Derr (1972, p. 113).

17. See, for example, March (1965) and Dubin (1974).

18. The complete questionnaire appears in Schmuck, Runkel, Saturen, Martell, and Derr (1972, pp. 261-263).

19. The exercises we use would take too much space to reproduce in this monograph, so we refer the reader to Schmuck, Runkel, Saturen, Martell and Derr (1972, pp. 270-280); the Improving Teaching Competencies Program Preparing Educational Training Consultants (PETC) I training system; and Pfeiffer and Jones, for examples.

20. This matrix is adapted from one by Wallen (1972).

This chapter describes a number of "rules of thumb" and presents an even larger number of things to think about. In many cases we have described choice-points where alternative strategies have different and possibly preferable consequences. The guidelines are arranged to illustrate four interdependent sets of behavior that every change facilitator must exhibit: (1) building and maintaining a relationship with the system; (2) knowing everything relevant about the system and changes that are brought about; (3) tailoring interventions to fit the system that is to be changed; and (4) growing and changing as the system does. The chapter ends with a case study of what one person did to facilitate system change in his elementary school.

Chapter 4

GUIDELINES FOR FACILITATING CHANGE

More than with the other chapters, we debated about the title of this one. Since "Advice" sounded too presumptuous and "Issues" too forbidding, we settled on "Guidelines" as the most inclusive and humble rubric. But some of the ideas in the chapter are intended as advice. For instance, from our own experience and from reading what others have found, we do not hesitate a minute to say, "Don't do it alone if you can get a partner." Other ideas, such as "Don't let your personal preferences for certain facilitator functions get in the way," are not advice so much as they are things to think about. We hope both kinds of ideas are useful.

One definite problem that any writer would face

with a chapter such as this is the one of defining the audience specifically. We refer the reader back to Chapter II for the discussion of the various roles that facilitators of change can perform and for our comments on how "insidedness" and bases of influence can limit or expand the would-be facilitator's role options.

Many of the ideas for this chapter first appeared in an article we wrote with our friend and colleague, Richard Schmuck (Schmuck, Arends, and Arends, 1975). Although this article, "Tailoring Consultation in Organization Development for Particular Schools," was written for facilitators who are not a part of the system they help, we think many of the ideas are quite generalizable to those in other situations. Other resources for this chapter include a short paper by Shepard (undated), the *Handbook of Organization Development in Schools* (Schmuck, Runkel, Saturen, Martell, and Derr, 1972), and *The Change Agents Guide to Innovation in Education* (Havelock, 1973).

THE BASIC MODEL

The guidelines in this chapter illustrate a four-part model that includes the following points.

1. Build and maintain a relationship with the system so you can facilitate change.
2. Know everything you can about the system and the changes you bring about.
3. Tailor your change efforts to what you know about the system.
4. Grow and change as the system does.

Although the guidelines in this chapter are presented to illustrate one or another of these four points, relationships among them are very important. For example, your relationship with the system largely will determine what you can know about it. Persons in high authority positions may never find out how others view them, yet they have an easy time understanding external pressures on the system. Persons in roles that are more insulated by the system may never understand problems at the system's boundary, but they have little trouble defining system norms and the day-to-day behavior they support. Or, as another example of interdependence, the notion of tailoring an intervention is built on the assumption that you *have* information before you begin and *continue* to elicit information as you proceed to make possible further tailoring. Finally, information about the consequences of your actions can enable you to grow; but as you change, so will the relationship you have with the system. In short, progress in one set of tasks will make other sets easier. In exchange, however, a problem in one set of tasks—in your relationship with a key person or with inaccurate information about one thing—can plague you in all others.

BUILDING A RELATIONSHIP

Many of the issues in building a relationship involve themes like trust, dependability, and rapport between you and others in the system. Do they trust your motives? Are they expecting to do some of the work or are they depending on you to work magic for them? Is there enough rapport among you to join in a collaborative effort? Others may or may not state these questions explicitly. If they do, you ought to be prepared to share your ideas and feelings. If no one else initiates the discussion, you must. Clarity about these issues is of prime importance.

In addition to sharing information about the role you will assume, it is important to see that targets for change are discussed and negotiated. Candor and openness about where you want the system to go may help others to share their preferences. Exhibiting receptivity to others' ideas may model one of the goals you want to pursue. It is a lot easier to know the goals and disagree on them than to leave individual goals and agendas under the table. Disagreements can be treated as fodder for problem solving, and nothing will be gained ultimately by denying problems or pretending that they do not exist.

Establishing rapport does not necessarily mean that you have to be like everyone else in the system. It is possible to facilitate change if you are different from others in terms of sex, race, values, or political persuasion. It does become more difficult as the number of differences increases, but the most severe difficulties arise when differences are ignored. It is important to discuss anything that may provoke curiosity, conflict, or dependency later on, so that resolutions can be sought early.

As "outside" consultants to school organizations in the past, we have found that frequently asked questions include, "How long will it take to change?" and "What do you (the consultants) expect of us?" We think it is helpful to explain not once but many times what we expect and hope to gain. Our answers about the cost in terms of time are always as realistic as possible, but we do not hesitate to say that we do not know or would like to negotiate a more definite agreement after a few weeks. Since both the change agent and the system can be hurt by unfulfilled promises, it is better to keep the relationship on a short-term and experimental basis until all feel comfortable with more definite and long-term commitments.

Meaningful relationships are hard to build with words alone. You first may want to ask for permission to demonstrate what you can do and to let others experience what that would be like. Others who know your skills and understandings might help you plan, carry out, or reflect on such a demonstration event. Handouts or graphic displays can also supplement or reinforce oral messages.

Building and maintaining relationships with key authorities or gatekeepers in the system is particularly important. If you do not hold such a position, your very first task may be to talk to those who do. If you prefer, concentrate on building support from enough of your peers and subordinates to convince those in authority that you are the one to do the job. Remember that, although your supervisors may not have the motivation or skill to bring about changes themselves, they can often squelch your effort single-handedly. They certainly should not be left out of discussions.

Although permission from the "top" of the organizational hierarchy is necessary, it is not sufficient. If you want to change the system, all of its members must recognize their needs and be willing to receive help. When and as long as they do, and when and as long as you can and want to help, the relationship has potential for mutual growth and exchange.

Knowing About the System

Different people use different words to describe the process of coming to know about a system. Diagnosis, assessment, evaluation, and research all involve activities that include matching information or data about the system against criteria, measurement scales, questions, or hypotheses. Since there is no way to point out the finer distinctions among these terms in a monograph this short, we will assume that readers have at least some notion of what we are talking about.

Potential information that you might collect about a system is endless in both amount and kind. For instance, a person with linguistic orientations might study word usage for years, or a political scientist might examine structures and policies over decades. Because we focus so definitely on changes in subsystem processes as the way to change a system, we recommend that data collection efforts concentrate in that domain. For example, instead of initially studying the content of problems by gathering information about past trends or projections for the future, examine how group members solve or do not solve problems. Or, instead of listening to what group members talk about, concentrate on their use of communication skills or how they fulfill communication functions in the group. Although you may wish to draw distinctions among the ways they solve different kinds of problems or communicate about different topics, look first at processes that will be the target of whatever intervention you make.

Many of the tools in Chapter III can help in collecting information about the system, and virtually all of them focus on a single process such as conflict management or meetings. Several of them would have to be used before a comprehensive picture could be drawn. Beginners at process observation, interviewing, or questionnaire design and administration may prefer to focus on one process at a time, but information about all processes and the relationships among them is eventually necessary.

It is important to use multiple strategies for collecting information. Before reaching generalizations about the way conflict is managed, for example, you might observe in several meetings, interview parties to a conflict, and administer a questionnaire to everyone. A combination of formal and informal procedures is more effective than either one alone, and collecting information from various people with different perspectives on the same issue is better than relying on only a few sets of eyes and ears.

We have frequently recommended that you summarize your observations or information for the group. Feedback to them from you can serve several purposes: (1) it can help you verify the accuracy of information you have collected, (2) it may stimulate the group to do some problem solving about the situation you have observed, (3) it will provide a setting in which you can gather further information about the group, and (4) the act of collecting data about processes and putting them before the group can provide a valuable model for others to follow.

Do not be suprised if system members react negatively to your first efforts at data collection and feedback. Some systems do not necessarily like to be diagnosed or watched or asked questions. Avoiding jargon and explaining what you want to do and why you want to do it in advance may help to alleviate some suspicion and resentment. But do not let others talk you out of this effort; insist on having information that will guide your actions and help you to understand the consequences of actions that you have already taken. If people insist that they already know what the problems are and do not want you to interfere with their need for action, focus data collection on their perceptions of why the problems have not been solved previously or on the consequences of leaving them unsolved.

Remember that data collection is a never-ending

task. Unlike the medical model in which a diagnosis is completed before the prescription is written, learning about a system will continue as long as the effort to change it. Repeated use of the same observation or interview schedule can provide valuable information about the process of intervention and change as well as about the group processes on which it focuses.

TAILORING INTERVENTIONS

By describing the multiple functions that change facilitators can fulfill and by presenting multiple tools that can be used to intervene in the system's functioning, we hope that we have made the point that there is no one way to change a system. The most important guideline of all is to create a strategy that starts where the system is and moves it toward agreed on goals. The act of designing a tailor-made strategy is most challenging and important.

Starting where the system is may not be as easy as it sounds. You may decide that certain prerequisites have not been met and that there is a need to increase the system's readiness to receive the kind of help you can or want to give. For example, we have found that success of an organization development intervention is integrally related to the presence of certain norms when the intervention begins.[1] If the group we want to work with does not expect its members to work together collaboratively, or if it does not reward them for doing so, we first try to help create this norm. We will also train people in ways to express their differences if this behavior is not common. We are always prepared to increase readiness for accepting our consultation if this does not exist.

Building system readiness to profit from the functions that you want to perform may provoke you into examining yourself. If you are convinced that process consultation would be best, for instance (or if it is the only function that you can skillfully perform), you may have to prepare others to accept and profit from that. On the other hand, you may find it easier to increase your knowledge and skill in performing the convener function if others are ready and willing to put you in that role. In either case, do not let your personal preferences for performing certain facilitator functions get in the way of seeing what the system needs and is willing to accept.

A well-tailored intervention includes continuous strategies to maintain your relationship to the system and to collect additional information from it. Once you have intervened, talk to others about what you did that was helpful and unhelpful, and establish new contractual agreements about what you will do the next time. Documenting what happened when you intervened in a certain way will provide a data base that you can use in future planning.

It is possible to create more than one design that is tailored to the strengths and weaknesses of the system. One strategy is to start on the process that needs the most attention. If you follow this strategy with a group that has good meetings, resolves conflict, and is comfortable with the way decisions are made, you might begin with the communication process, then move to goal setting and problem solving; introduce additional tools for meetings, conflict management, or decision making only as needed. Likewise, if you think an ineffective decision-making process is the most severe problem, you could teach people alternative decision-making procedures and then help them to explore the consequences of each procedure of effective communication and conflict management. This strategy can also be implemented by working first with the people who are least accepting of you and your efforts.

The other popular strategy and the one we personally prefer is to capitalize on more effective processes to strengthen less effective ones. With this strategy you would help effective decision makers decide the kind of communication process to use, help good goal setters agree on the kind of meetings they would like to have, and first work with the subgroup that is most supportive of your effort. But neither strategy has definite advantages over the other, and some combination may be most effective.

The sequence with which subsystem processes are improved is less important than keeping them all in mind. An improvement in communication that encourages more people to say what they think may make previously adequate decision-making procedures obsolete, for instance. A newly identified goal may surface conflicts that cannot be handled with present norms. Consideration of all processes and the relationships among them requires the change facilitator to "light many fires" simultaneously and to help people tend them all.

GROWING AS THE SYSTEM CHANGES

If you are successful in facilitating change in the system to which you belong, new norms and roles will affect you as well as others. More than that, if you put your whole self into the effort, you deserve something in return. This is not a selfish stance, but one that can insure your "survival" in an effort that you do not wish to be your last.

Personal growth and change is precipitated by feedback. Knowing how others perceive you and feel about your behaviors activates personal problem solving to bring their impressions more in line with your goals. If people see you as less knowledgeable or skillful than you would like to be, for instance, you can start to resolve the discrepancy by increasing your knowledge or skill.

But others must be motivated and competent to give you feedback for this purpose. Sometimes everyone in the system you work with will do this for you, but more often than not you will have to rely on a few colleagues for a support system. It is for this reason that we counsel so strongly for teaming up with someone else who wants to facilitate change. A partner or partners with whom you can verify information, plan the strategy, and debrief what happened is a most valuable asset to any facilitator. This is particularly true when the going gets tough and you need "warm fuzzies" or a nudge to start your personal problem-solving.

Shepard's list of "rules" (undated) begins with "Stay alive." He says that this means (1) taking risks on purpose as part of your strategy rather than by accident, but not if they will not lead toward your goal; (2) putting your whole self into the effort instead of stretching yourself with many efforts or halfheartedly trying to do something you really want very much, and (3) loving yourself instead of being self-disparaging, thereby suppressing your potential and wasting your energy in defensive maneuvering. Partners can help you to live by this rule.

We do not want to conclude this section without mentioning one final guideline that we think has helped us as group facilitators of system change. It is the three-word guideline we followed to write this monograph: read and write. There is a wealth of information already available, and new articles and books come out almost daily. One criterion we used for selecting books for the bibliography was that they can start you searching for more information in almost any direction you want to go.

But all the books you can possibly read and all the advanced training you can get cannot tell you everything you might want to know because not everything is known. Changing systems is relatively newer than changing individuals, and many strategies are unexplored. Writing about what you have done may give others just the idea they are looking for. Taking the risk to report what you have learned—even if you learned it the hard way by failing more than once—can help all of us who care about system change to do a better job.

MARK'S STORY

This case study describes the efforts of a teacher and part-time counselor (we have called him Mark, although that was not his name) who acted as a facilitator of system change to his own elementary school faculty. Although Mark was not a full-time counselor or school psychologist—his small school district could not afford such luxuries—his case does illustrate some of the unique difficulties encountered by a "very internal" consultant, describes some of the problems of the solo consultant, and portrays a common education consulting situation. We include it here to help counselors and school psychologists see how the guidelines outlined in this chapter can be applied by someone other than an outsider.

We first met Mark when he was a participant in a special training program called *PETC-III*.[2] Mark had much invested in education and in human relations training. Besides his B.A. in Education, he had had training in race relations work in the archdiocese and had taken many NWREL instructional systems, including all the *PETC-III* prerequisites.[3]

Mark felt confident about his abilities and clear about what he wanted to do. He was sure from the beginning that he wanted to work with his own school and that he wanted to work alone. He said, "The staff already uses me as a consultant for some kinds of things. I guess I'm kind of a loner, but I don't think they (the staff) would want an outsider hanging around. It's more realistic for me to work alone."

Mark's seventh- and eighth-grade school was one of two elementary schools in an inner-city parochial school system. Both schools gave special attention to children who were below grade level in basic skills—usually children with learning difficulties or behavior problems. Each school was small, with about 200 students; and each class was also small, with 15 to 17 students.

At one time, there had been four elementary schools in the district. As attendance dropped, the archdiocese had decided to close two of the schools, leaving two neighborhood schools—one for grades 1 to 6, and one for grades 7 and 8.

Mark was one of the 12 teachers and the only part-time counselor in the seventh- and eighth-grade school. He had been involved in planning the new program when the school opened and in trying to establish norms of trust, cooperation, growing self-respect, and respect for others. Mark described himself as "like the other teachers. I enjoy working on my own, but I'm very committed to the school and the kids."

Entry

Mark chose an informal entry style because, as he said, "An internal person in a small system has to be informal and be himself—not play the role of a 'big expert.'" Mark also reported that he had the advantage of an expert and referent power base from which to work. "The staff knew that I had had some training and was getting more. They had rallied around me in the past and I knew they would do so again if I asked them to."

The first thing Mark did after deciding that he

wanted to work in his own school was to ask his colleagues if he could work with them on building trust
and openness in the faculty. After the staff informally
agreed to undertake the project, he approached the
principal and said that he would like to do this project
with the staff. They discussed the project and whether
Mark should be perceived primarily as a consultant, a
teacher, a counselor, or a "human being." Then Mark
went back to the staff and told them he had the principal's approval. The staff agreed to participate in the project.

Mark used a formal contract form because these
were required by *PETC-III,* although he thought the
forms were somewhat inappropriate for an internal consultant in a small system. Other staff members reacted
with comments like, "You want us to sign a contract?"
but agreed to do so because they liked the idea of
being specific about agreements they had with Mark.

Only later did Mark discover that the formal contract did not help him to discover the principal's discomfort and uncertainty about dealing with feelings. Although she was in agreement with the project, it seemed
to Mark that she did not have a clear picture of what
might happen.

Diagnosis

When Mark first thought about his system, he described
the teachers as "strongly opinionated individuals with
definite ideas." He also thought that the faculty lacked a
basic teamwork capability. He wrote in May, "Combining
people's efforts in an organization necessitates understanding each other, agreeing to norms of influence, ar-
riving at school commitments in decision making, and
coordinating their efforts."

Mark's long-term goal was to deal with his own system's needs for human maintenance, problem solving, and decision making. He also wanted to avoid "laying stuff on people and being the expert." He wanted to share the diagnostic data so that the group would own it.

His diagnosis was based on data he collected with a "needs assessment" questionnaire during a faculty meeting in May. Eight faculty members responded to 17 statements[4] such as, "Ask others who seem upset to express their feelings directly," and, "Use formal voting as a way of making decisions," by indicating whether group members did or did not, should or should not exhibit those behaviors. When everyone had completed the questionnaire, Mark tallied all responses on a large chart and asked others to consider discrepancies between what staff members were doing and what they should be doing.

Then he asked the staff to generate a list of group goals and to indicate which ones they were most committed to. By comparing the list of goals with data from the questionnaire, the staff then generated a list of major problems to attack in the fall. Mark and the staff felt good about this meeting. As Mark said later, "We were well on our way towards 'the year of good feelings.' "

Interventions

During the summer, Mark began to plan for a two-day workshop with the staff just before school started. In a report, he wrote the following:

> My goals are: (a) to increase the adequacy of problem solving in the staff, (b) to increase the accuracy with which we understand each other's feelings about the organization, and (c) to increase the adequacy with which we manage the school. My basic *assumption* is that if the faculty can identify the real issues that keep us from working well together, we must (a) have team skills, (b)

understand our values, and (c) analyze our purpose for working together.

The *strategy* that I'll use will be a process link, presenting skills so we can uncover and resolve the real issues in our group.

I hypothesize that the *outcomes* will be a process that will enable us to deal with larger organizational issues.

He planned the two-day workshop to focus on team building. He wanted to help the staff explore the ways in which humanistic values and needs for autonomy and achievement—typical qualities of faculty members—inhibited or facilitated their ability to work together. In Mark's words, "I wanted to build on group concepts and to focus on team-building issues. If we created more sense of teamness, we'd function better and take care of our human needs, too."

The workshop included a review of problems and goals identified the previous May, as well as a "brushup course" on interpersonal communication skills and constructive feedback. During the two days, the faculty spent most of the time examining roles and norms in order to be more explicit about their expectations of each other. They also did a force field analysis on their goals of increased supportiveness in the staff and favorable student attitudes toward work, self, and others.

As "Forces For," the staff listed: (1) we are a small, friendly group; (2) we have interpersonal communication skills; (3) we have common interests in teaching; and (4) all of us are sincere, desire new ideas, and are open-minded. As "Forces Against," the staff listed: (1) we fear knowing each other too personally and hurting each other's feelings: (2) we lack time and sometimes put efficiency ahead of good communication; (3) we are not always aware of each other's philosophies and values; and (4) we sometimes work in competitive ways, judging and making assumptions about each other's behaviors.

A major decision during the two days was that the faculty would meet once or twice a month during the school year to discuss group maintenance issues. The staff hoped that these meetings would help to maintain the following norms: (1) to give and receive feedback at all appropriate times; (2) to work as a team to monitor each other's behavior; and (3) to rotate responsibility for convening these meetings so that everyone "owned" the OD effort.

According to Mark, the staff also spent a great deal of time exploring inclusion questions like "What is Mark's position as counselor-teacher-change agent in this faculty?" and values questions like "What do other faculty members believe in strongly?" He played a key leadership role in the workshop, but repeatedly told the staff, "I'll need help. I don't want to be isolated and I hope you don't expect me to be perfect. Give me the freedom *not* to be perfect as an internal guy."

One unexpected and pleasant result of the workshop for Mark was that he gained an informal "partner," Elaine, another teacher who had received training in communication and group process. He said, "She would pick up stuff I missed and give me feedback about how I did. She was also someone I could sit down with over a beer and talk to about how things were going, and there were times I really needed that!"

Another result of the September intervention was that teachers were more open and relaxed and came to talk to Mark about problems. The staff continued the monthly maintenance meetings, although they were not always held on a regular basis because of the pressure of other tasks.

During the fall, problems in the interaction between Mark's school and the other school in the district began to surface. The teachers in Mark's school asked Mark to help them work with teachers from the other school. Because the other school had an autocratic principal and a more rigid philosophy, there were problems, and those problems had an impact on the seventh- and eighth-grade school (Mark's) it fed into. Teachers in Mark's school were concerned about what Mark called "the typical hassles over philosophical values and directions."

Mark did his second intervention in December when he convened a meeting that he dubbed "The Red Purge." Teachers from both schools spent a day discussing how students should be taught and ranking characteristics of each other as teachers. Mark assumed that these activities would uncover values, build trust, and open up feelings. Both faculties and Mark thought the meeting was helpful and "cathartic."

As his final intervention in May, Mark held a half-day session with his own staff. This meeting was an opportunity for the staff to take a look at what they had been doing, to share feelings about the year's OD effort, to sum up their efforts, and to get closure on the project. This, too, was a well-received meeting and, according to Mark, "left people feeling good."

Outcomes of Consultation

Mark did not try to do any formal evaluation because he thought that it was awkward for an internal person to attempt that. At the end of each intervention, he did ask staff members for specific activites that they saw as helpful and unhelpful. He also used feedback from Elaine to check on how he was doing.

In general, Mark felt good about accomplishing what he set out to do. He thought that focusing on team building and group process skills had been a manageable project for him, and that the staff had come closer together and operated better—at least for a while. According to Mark, "The staff was pleased, I think. Although no one said it explicitly, I know they wanted me to continue in the consultant role. All around, I have to call that year 'The Year of Good Feelings.' "

Sometimes during his project, Mark had wondered if he was doing what he meant to do and whether his interventions had enough impact to be important. However, Mark's biggest problem was that he ran out of energy. Afer that year, he did not want to deal with others' stresses any longer. He now describes the major difficulty of internal consultation: "You get yourself pegged in this role and you can't get out of it. The internal consultant is always available to be sucked into every difficulty. When you are an internal consultant, you can't ever leave that place. Sometimes I felt like George Washington during his second term—I just wanted to retire to my farm and till the soil with my kids in the classroom. One of the things I failed to do was train others to take the responsibility for making the project keep happening."

Comments on Mark's Story

In many ways, Mark had a "success experience" as a facilitator of system change: (1) he managed to find a group of people who wanted him to help them improve their organizational functioning; (2) he diagnosed—with their help—the major organizational problems that they were facing; and (3) he conducted three major interventions that he and his clients evaluated favorably.

His story is also one of a solo consultant who "burned out" and who had trouble freeing himself from the role when he wanted to stop. The story suggests two important questions:

1. Why was Mark as successful as he was?
   -Was it because he did not start entry from the

"top" of the system, but instead went straight to those he wanted to help and then got the principal's approval?
  -Was it because his clients attributed expert and referent power to him while seeing him as "one of the folks"?
  -Was it because he relied heavily on informal methods of diagnosis and intervention?
  -Was it because he found a partner who could give him support and feedback?
2. Why did Mark "burn out"?
  -Was it because he worked alone and took on too big a task?
  -Was it because his "clients" never distinguished "Mark the teacher" from "Mark the consultant" in spite of his efforts?
  -Was it because other *PETC-III* trainees could not empathize with his case or because the system includes too little that is relevant for consultants as "internal" as Mark?

Exploration of questions like these should help readers of this monograph look at what they might want to do and want to avoid.

SUMMARY

Previous chapters have delineated specific albeit various aspects of change strategies. This chapter brings these aspects together and presents a four-part model for implementing change: (1) building a relationship with the system to be changed; (2) learning as much as possible about the system and the changes sought; (3) tailoring change efforts to meet the needs of the system; and (4) growing with the system. A case study is presented that uses the model in a small elementary school where an "internal" facilitator (counselor) offers to work solo with the remaining faculty to open communication, establish change goals, explore and implement possible change strategies, evaluate successes and failures, and establish maintenance procedures. Although this attempt was successful, it had drawbacks for the facilitator that are offered for consideration to those who may assume a facilitator role.

NOTES

1. Many interesting findings about organizational readiness appear in Runkel, Wyant, and Bell (1975).

2. *Preparing Educational Training Consultants (PETC-III)* is the last in a series of instructional systems developed by the Improving Teaching Competencies Program of the Northwest Regional Educational Laboratory (NWREL). *PETC-III* provides participants with the opportunity to acquire knowledge, skills, and sensitivities that constitute a change process termed organization development (OD).

Participation in *PETC-III* training lasts over an eight-month period during which *PETC-III* trainees complete a one-day pre-workshop assignment, attend 17 days of workshop meetings, and spend approximately ten days doing a practicum. The practicum consists of conducting an organization development project that focuses on improving the functioning of a client organization.

PETC-III trainees learn of the following stages of OD consultation: (1) developing a need for change, (2) establishing a client consultant relationship, (3) diagnosing the client's problems, (4) examining alternatives, (5) intervening and transforming change efforts, (6) stabilizing change efforts, and (7) terminating the relationship. We have used these stages or organize this case study.

Mark's story appears in a monograph we have written with Mary Ann Smith and William Ward. This monograph will include case studies of five trainee teams (Mark was the only "solo") that participated in the followup study of *PETC-III*. His story is somewhat abbreviated here, so readers interested in more detail or other case studies should refer to Arends, Ward, Smith, and Arends (1977).

3. See the annotated references for training systems of the Improving Teaching Competencies for names of these prerequisite instructional systems.

4. The 17 statements Mark used were: (1) ask others who seem upset to express their feelings directly; (2) tell colleagues what you really think of their work; (3) avoid disagreement and conflict

whenever possible; (4) be concerned about other people's problems; (5) only make a decision after everyone's ideas have been heard; (6) push for new ideas, even if they are vague or unusual; (7) ask others to tell you what they really think about your work; (8) keep your real thoughts and reactions to yourself by and large; (9) be skeptical about things; (10) point out other people's mistakes, to improve working effectiveness; (11) try out new ways of doing things, even if it is uncertain how they will work out; (12) stay cool—keep your distance from others; (13) use formal voting as a way of making decisions in small groups; (14) spend time in meetings on emotional matters that are not strictly germane to the task; (15) be critical toward unusual or "way out" ideas; (16) stick with familiar ways of doing things in one's work; and (17) trust others to be helpful in difficult situations.

This final chapter restates the basic premises from which the authors began. These include: (1) that school systems do, need to, and can change; (2) that there are ways to promote desirable and lasting changes; and (3) that counselors and school psychologists can be involved in this process and work toward purposeful change in their own settings. Some of the implications of these basic premises are discussed; some of the future needs and unresolved problems are described.

Chapter 5

TRENDS FOR THE FUTURE

THE ROLES OF COUNSELORS AND SCHOOL PSYCHOLOGISTS WILL CHANGE

In Chapter I, we reviewed predictions by Singer, Whiton, and Fried (1970), Murray and Schmuck (1972), Medway (1975), and Meyers (1973), that roles of those in the counseling and psychological services will include a systemwide change orientation. Others have provided evidence that this kind of role change is happening or needs to happen (see Gallassich, 1974; Hayman, 1974; Lee, 1972; Kessinger, 1975; and Schmuck and Schmuck, 1974).

We agree with the fact that role change in this direction needs to happen. As an example of why we think this way, let us quote a school psychologist we interviewed.

Us: How are things going for you this year in the district?

The Psychologist: You won't believe how bad they are! I'm responsible for taking referrals from 18 schools and I get more than 60 a month. In 20 working days I can't do anything more than put band-aids on people and situations that need radical surgery. On top of that, the director of our division is retiring and all of us at the other end of the stick can't get any help.

Us: What would you like to be doing instead? What kind of strategy for a school psychologist like you would be worthwhile?

The Psychologist: I think the only hope for kids is if everyone on the district payroll pitches in to help. But that's going to take some systemwide changes. It's not going to help to train teachers to administer psychological tests or to train secretaries to schedule students into classes so counselors and school psychologists have more free time. Instead, we all need help to sit down and talk to each other so we can figure out ways to solve these problems of helping kids that need so much. Lots of teachers and administrators could do that kind of helping with no additional training, but until we sit down and agree we should do it . . . until the criteria for evaluating teachers include a responsibility for providing the "classical" psychological services . . . .

Us: What can you—all by yourself—do to bring about that kind of system change?

The Psychologist: A few months ago, I would have told you there was nothing I could do. Now I don't know. I have been talking to other school psychologists, counselors, and social workers about this mess. They agree with me that we might as well stay home for all the good that one-on-one "band-aids" do. We're still talking about what to do. This week a few of us went to the superintendent to say we wanted to help select our new boss. He surprised us by saying the division could have two representatives on the five-person selection committee.

> We surprised him, I guess, by coming in. He told us that he hadn't known we cared. Well, we showed him and now we can try to make sure that the new director is sympathetic to our plight.
>
> Us: What do you think your strategy will accomplish?
>
> The Psychologist: I don't know for sure. Already I have reached out for others who share my dilemma and feel a whole lot less lonely. Down the pike, I hope our division can set the example of a group that solves important problems its members have. If we can help each other, maybe we can learn the skills to help other groups.

The other psychologists and counselors we talked to all expressed similar frustrations with their current roles. All expressed the hope that by joining together they could make significant changes in the systems of which they were a part. None of them had more specific ideas about how to proceed, but they all insisted that they could and would find a way.

In various ways, these individuals were trying to change their roles so that they could act as facilitators of system change. Their enthusiasm and sincerity convinced us that roles of counselors and school psychologists will change—although it may take time.

PROFESSIONAL GROWTH OPPORTUNITIES WILL EXPAND

A second trend we see is that counselors and school psychologists will find more professional growth opportunities in the area of system change. More books and articles will be published, and the chances to belong to professional organizations will be heightened as new networks of technical assistants interested in system change are created. In addition, more programs to teach skills and techniques that facilitate group and organizational change will become available.

An example of the type of training program we envision has been developed by the Improving Teaching Competencies Program of the Northwest Regional Educational Laboratory (see bibliography). Some of the training systems from this program focus specifically on the human relationships and group processes of classrooms and schools. They include topics such as interpersonal communication, group problem solving, interpersonal influence, and social conflict. Other training systems in this series, those called *Preparing Educational Training Consultants (PETC),* prepare persons in the educational community to assume new roles similar to those we have defined for counselors and school psychologists in this monograph. Trainees learn how to help others set goals, clarify communication, reach out for relevant resources, solve problems, make decisions, and cope with interdependence and conflict. By involving themselves in a series of workshop and practicum experiences over a two-year period, trainees learn to work with small groups and subsystems of school organizations.

Mark, the internal consultant whose story appeared in the previous chapter, is just one of hundreds of teachers, counselors, administrators, and school psychologists we know who have gained these skills.

NEW SUBSYSTEMS FOR ORGANIZATIONAL RENEWAL WILL BE CREATED

In the near future, if present trends prevail, counselors and school psychologists interested in facilitating system change will no longer have to operate in isolation or create their own support systems. Instead, we think school districts and other educational agencies will find ways to attend systematically to the skills and processes needed for effective organizational functioning. These efforts we believe, will be institutionalized in the form of new organizational structures.

Few districts at the present time have this institutionalized capability, but several experiments by the Program on Strategies of Organizational Change (in the University of Oregon's Center for Educational Policy and Management) have shown the potential of this idea. Consultants in this program have trained and studied cadres of group and organizational process consultants in Kent, Washington and in Eugene, Oregon. Cadre members perform training and consulting functions for school staffs, teaching or administrative teams, and other work groups in the district. The success and staying power of these cadres may be attributed to four key factors (Arends and Phelps, 1973; and Runkel, Wyant, and Bell, 1975).

1. Members of cadres are drawn from various roles and sybsystems within the district on a volunteer basis. During the regular working day, they are counselors, teachers, principals, and school psychologists. They do not consult or try to help work groups of which they are a part, but instead work upon request and through released time with other groups and subsystems in the district.

2. Cadre members are organized and work in teams rather than alone. This teaming arrangement allows a natural support system for members, insures a range of resources and skills upon which to draw, and provides a setting for continuous learning and growth. New teams are formed to work with each new client subsystem.

3. Cadre members do not give advice about the content of their clients' problems. They do not pose as experts in curriculum, finance, classroom management, or whatever. Instead, they specialize in improving group and organizational processes with strategies similar to

those described in Chapter III.

4. Cadre members regularly pay attention to their own task accomplishment and group process problems in regular meetings. Consultants from outside the district are occasionally invited to provide additional training or consultation so that this subsystem can be a model of effective functioning.

In both Kent and Eugene, counselors and school psychologists have played an active role in these cadres. Their knowledge and training in school psychology equipped many with frameworks to see the value of this type of service. Many adapted skills they had used in traditional counseling to consultation in group and organizational settings.

We believe the creation of new subsystems for organizational renewal holds promise and will be more widely tried in the near future. Counselors and school psychologists will be called on to participate in and facilitate this change.

SUMMARY

This chapter synthesizes some of the ways in which school helping personnel can and do effect change within their schools. It presents an interview with a school psychologist who first expresses concern over his inability to make a difference in his work setting, but who comes to realize that he has already sowed the seeds for change.

Future trends are viewed as holding great potential for professional growth among those who are interested in becoming facilitators. Systems will become more responsive to the need to provide and train cadres of change agents whose primary function will be to improve group and organizational processes.

Probably many other changes in the business of

changing educational systems will come to pass in the next decades. Experimentation and research continue as many more people put their energies into such efforts. We hope this monograph has convinced old pros and neophytes alike that the benefits are worth the risks. Hopefully our joining together and reaching out to each other will give us all the thrill of work well done.

ANNOTATED BIBLIOGRAPHY

Arends, R. I., Phelps, J. H. *Establishing organizational specialists within school districts.* Eugene: University of Oregon, CEPM, 1973.

Discusses in detail the entry, training, and maintenance procedures used in two school districts to establish internal cadres of organizational specialists. Appendices include the training design for the cadre and a "report card" of interventions done by these cadres.

Arends, R. I., Arends, J. H., Schmuck, R. A. *Organization development: Building human systems in schools.* Eugene: University of Oregon, CEPM, 1973.

A booklet to introduce the theory and strategy of OD. Includes information that school personnel typically want as they consider their own desires to enter into a change effort. Can be accompanied by a slide-tape show of the same name.

Argyris, C. *Intervention theory and method: A behavioral science view.* Menlo Park, CA: Addison-Wesley, 1970.

Excellent presentation of theory and strategies to guide consultants who work and intervene in organizations.

Most examples are from industrial organizations, but applications to school settings are easily inferred.

Fordyce, J. K., & Weil, R. *Managing with people: A manager's handbook of organization development.* Reading, MA: Addison-Wesley, 1971.

Concentrates on the joint management of change and presents particular methods that have proven useful in this process. Ideas for the experienced and the neophyte change agent alike.

Fox, R., et al. *Diagnosing the professional climate of schools.* Washington, D.C.: NTL Learning Resource Corporation, 1973.

Explains numerous techniques for diagnosing the norms and roles of a school staff. Sample questionnaires and interview schedules are included.

Gordon, T. *Teacher effectiveness training.* New York: Peter H. Wyden, 1974.

Includes information on communication and conflict in schools, presents a model for effective teacher-student (trainer-client) relationships, and features many case studies and sample dialogues.

Havelock, R. G. *The change agent's guide to innovation in education.* Englewood Cliffs, NJ: Educational Technology Publications, 1973.

Filled with guidelines, hints, and examples for those using planned change strategies to improve educational organizations. Appendices include many periodicals, books, agencies, and organizations in an annotated fashion.

Improving Teaching Competencies Program products. Portland, OR:
Northwest Regional Educational Laboratory.

Jung, C. C., et al. *Interpersonal communications.* Tuxedo,
NY: Xicom, Inc.
Emory, R. E. & Pino, R. F., *Interpersonal influence.*
Tuxedo, NY: Xicom, Inc.
Pino, R. F., & Emory, R. E. *Preparing educational training
consultants, I, II, III.* Portland, OR: Northwest Regional
Educational Laboratory.
Jung, C. C., Pino, R. F., & Emory, R. E. *Research utilizing
problem solving.* Portland, OR: Commercial Educational
Distributing Service.
Lohman, J. & Wilson, G. *Social conflict and negotiative prob-
lem solving.* Portland, OR: Northwest Regional Educa-
tional Laboratory.

A series of instructional systems for educators who wish
to understand or more effectively manage the group and
organizational processes in schools. Each includes materials and experiential-learning activities to be used in
workshop settings. Experience in workshops for several
products qualifies one as a trainer.

Koberg, D. & Bagnall, J. *The universal traveler: A soft-systems guide to
creativity, problem-solving, and the process of reaching goals.* Los Altos, CA:
William Kaufman, 1972.

In the tradition of soft-bound catalogues, this one uses a
variety of layouts, type styles, and information to describe
important features of the goal-setting, problem-solving,
and decision-making processes.

Lortie, D. C. *School teacher: A sociological study.* Chicago: University of
Chicago Press, 1975.

An interesting and provocative definition of the nature
and content of the ethos of the teaching professions.

Concludes with speculations on change that are based on a thorough understanding of schools as social systems.

Maier, N.R.F. *Problem-solving and creativity.* Belmont, CA: Brooks-Cole, 1970.

Brings together Maier's extensive research on individual and group problem solving. Excellent theoretical and research base for thinking about ways to improve the problem-solving effectiveness of school groups.

Miles, M. B. (Ed.) *Innovation in education.* New York: Teachers College, Columbia University, 1964.

Considers the theory, research, actual case studies, and principles that apply to innovation in education. Describes the American educational system as a setting for change in a way that is valuable both to researchers and practitioners.

Miles, M. B. *Learning to work in groups: A program guide for educational leaders.* New York: Bureau of Publications, Teachers College, Columbia University, 1959.

Although it is the oldest book in this list, it contains excellent suggestion and examples for group leaders on how to make groups more satisfying and effective.

Pfeiffer, J., & Jones, J. E. (Eds.) *The annual handbook for group facilitators.* LaSalle, CA: University Associates (yearly editions).

These books contain structured experiences for all kinds of change agents to use, instruments, lecturettes, and book reviews. Excellent sources for ideas of what can be done and how to do it.

Phelps, J. H. & Arends, R. I. *Helping parents and educators solve school problems together.* Eugene: Oregon University, CEPM, 1973

Describes the entry procedures, diagnostic techniques, and design characteristics of a change effort involving both parents and educators as clients. Particularly useful for explaining how helpers with varying degrees of insidedness can team to promote system change.

Sarason, S. B. *The culture of the school and the problem of change.* Boston: Allyn and Bacon, 1971.

An excellent explication of the importance of treating the school as a complex social system if meaningful change is to occur.

Schein, E. H. *Process consultation: Its role in organization development.* Reading, MA: Addison-Wesley, 1969.

An excellent and basic guide to the concepts and practices of process consultation. Includes guidelines and examples that apply to various subsystem processes and the relationships among them.

Schmuck, R. A. *Incorporating survey feedback in OD interventions.* Eugene: Oregon University, CEPM, 1973.

A thorough explication of principles to follow during formal data collection. Explains how the consultant gains insights and provides feedback to clients.

Schmuck, R. A., Arends, J. H. & Arends, R. I. Tailoring consultation in organization development for particular schools. *The School Psychology Digest,* 1974, 3, 29-40.

Provides a step-by-step approach for making entry, diagnosing systems, and designing interventions for schools.
The 16 guidelines treat the topics of Chapter IV in
greater detail.

Schmuck, R. A., & Miles, M. B. (Eds.) *Organization development in
schools.* LaJolla, CA: University Associates of LaJolla.

A collection of empirical studies of change projects using
organization development strategies. Several of the
studies include descriptions of entry, diagnosis, and design procedures.

Schmuck, R. A., Runkel, P. J., Saturen, S. L., Martell, R. T., and
Derr, C. B. *Handbook of organization development in schools.* Palo Alto,
CA: Mayfield Press, 1972.

The most complete source of theory and technology for
consultation using organization development strategies.
Spells out a theory of schools as social systems and includes questionnaires, exercises, and designs that can be
used. An expanded second edition by Schmuck, Runkel,
Arends, and Arends will be published by Mayfield Press
in 1977.

Schmuck, R. A., & Schmuck, P. A. *A humanistic psychology of education:
Making the school everybody's house.* Palo Alto, CA: National Press
Books, 1974.

Argues that schools can accomplish humanistic change by
changing the relationships between people. Chapter 8 describes the role that counselors might play in this process.
Excellent annotated bibliography at the end of each chapter.

Schmuck, R. A., & Schmuck, P. A. *Group processes in the classroom.* (2nd ed.) Dubuque, IA: Wm. C. Brown, 1975.

> Thorough review of group dynamics literature as it applies to the classroom as a learning group. Includes action ideas and hints for those who would intervene to improve classroom climates.

Simon, S., Howe, L, & Kirschenbaum, H. *Values clarification: A handbook of practical strategies for teachers and students:* New York: Hart, 1972.

> The plans and strategies to be used by teachers in their classrooms can easily be adapted by consultants for use in other settings.

Walton, R. E. *Interpersonal peacemaking: Confrontations and third-party consultation.* Reading, MA: Addison-Wesley, 1969.

> Another book in the Addison-Wesley series, this one provides guidelines and principles for those who help others surface and manage conflict.

Watson, G. (Ed.) *Change in school systems.* Washington, D.C.: National Training Laboratories, 1967.

> Includes descriptions of schools as social systems with particular properties, a strategy for changing school systems, and a description of the change-agent role within school systems.

REFERENCES

Arends, R. I., Ward, W., Smith, M. A. and Arends, J. H. *First time out: Case studies of five neophyte OD consultants.* Portland, OR: Northwest Regional Education Laboratory, 1977.

Benne, K., & Sheets, P. Functional roles of group members. *Journal of Social Issues*, 1948, 4, 41-49.

Berman, P., & McLaughlin, M. W. *Federal programs supporting educational change, Vol. IV: The findings in review.* Santa Monica, CA: Rand, 1975.

Dubin, R. *Human relations in administration.* Englewood Cliffs, NJ: Prentice Hall, 1974.

Fosmire, F. S-T-P problem solving. Eugene: University of Oregon, 1972 (unpublished manuscript).

French, J.R.P. & Raven, G. *The bases of social power.* In D. Cartwright (Ed.), *Studies in social power.* Ann Arbor: Michigan University, Institute for Social Research, 1970.

Gallessich, J. Organizational factors influencing consultation in schools. *The School Psychology Digest*, 1974, 3 (4), 40-44.

Gibb, J. R. Defensive communication. *Journal of Communication*, 1961, VXI, 141-148.

Goodlad, J. I., et al. *Behind the classroom door.* Worthington, OH: Charles A. Jones, 1970.

Goodlad, J. I. *The dynamics of educational change: Toward responsive schools.* New York: McGraw-Hill, 1975.

Havelock, R. G. *A guide to innovation in education.* Ann Arbor: Michigan University, Institute for Social Research, 1970.

Hayman, J. L. The systems approach and education. *Educational Forum*, 1974, **38**, 493-501.

Helmer, O. *The Delphi method for systematizing judgements about the future.* Los Angeles: California University, Institute of Government and Public Affairs, 1966.

Jacobs, A., & Spradlin, W. (Eds.) *The group as agent of change.* New York: Behavioral Publications, 1974.

Johnson, D. W. *Reaching out: Interpersonal effectiveness and self-actualization.* Englewood Cliffs, NJ: Prentice Hall, 1972.

Kessinger, P. Changing perspective of the roles of school counselor and school psychologist since 1950. Eugene, Oregon: University of Oregon 1975 (unpublished manuscript).

Lee, W. S. A new model for psychological services in educational systems. *Educational Technology,* 1972, June, 21-24.

Lippitt, P., Lippitt, R., & Eiseman, J. *Cross-age helping program; Orientation, training, and related materials.* Ann Arbor: Michigan University, Institute for Social Research, 1971.

Mager, R. F. *Goal analysis.* Belmont, CA: Lear Siegler, 1972.

March, J. G. (Ed.) *Handbook of organizations.* Chicago: Rand McNally, 1965.

Medway, F. J. A social psychological approach to internally based change in the schools. *Journal of School Psychology,* 1975, **13** (1), 19-27.

Meyers, J. A consultation model for school psychological services. *Journal of School Psychology,* 1973, **11** (1), 5-15.

Miles, M. B., & Lake, D. G. Self-renewal in school systems: A strategy for planned change. In Goodwin Watson, (Ed.), *Concepts for social change.* Washington, D.C.: National Training Laboratories, 1967, pp. 81-88.

Murray, D., & Schmuck, R. A. The counselor-consultant as a specialist in organization development. *Elementary School Counseling and Guidance Journal,* 1972, **7** (2), 99-104.

Runkel, P. J., Wyant, S. H., & Bell, W. E. Organizational specialists in a school district: Four years of innovation. Eugene: University of Oregon, CEPM, 1975 (mimeo).

Sarason, S. B. *The culture of the school and the problem of change.* Boston, MA: Allyn and Bacon, 1971.

Shepard, H. A. Rules of thumb for change agents. 1971 (unpublished manuscript).

Silberman, C. E. *Crisis in the classroom: The remaking of American education.* New York: Random House, 1970.

Singer. D. L., Whiton, M. B., & Fried, M. L. An alternative to traditional mental health services and consultation in schools: A social systems and group process approach. *Journal of School Psychology,* 1970, **8,** 172-179.

Wallen, J. *Charting the decision-making structure of an organization.* Portland, OR: Northwest Regional Educational Laboratory, 1970.

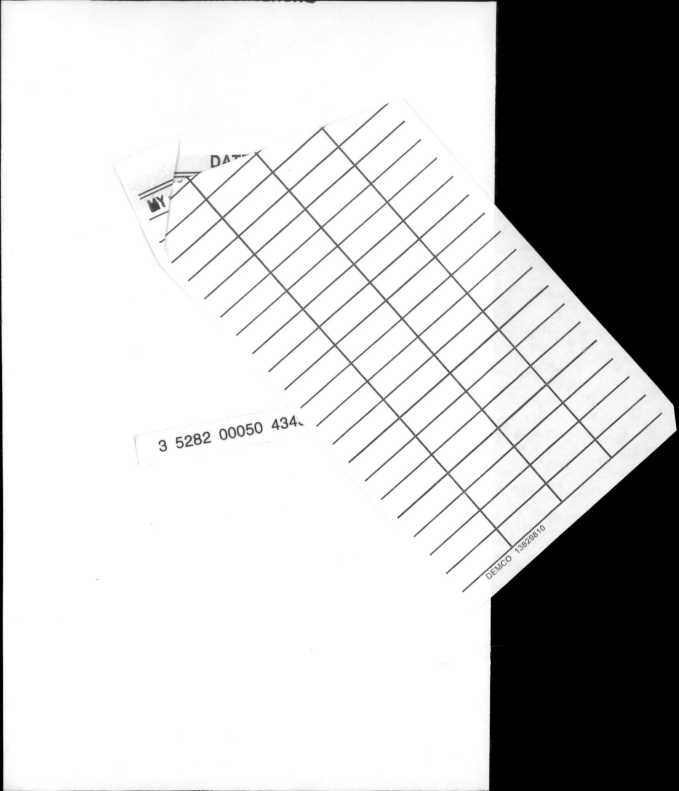

DEMCO 13829810